Practicing Geography

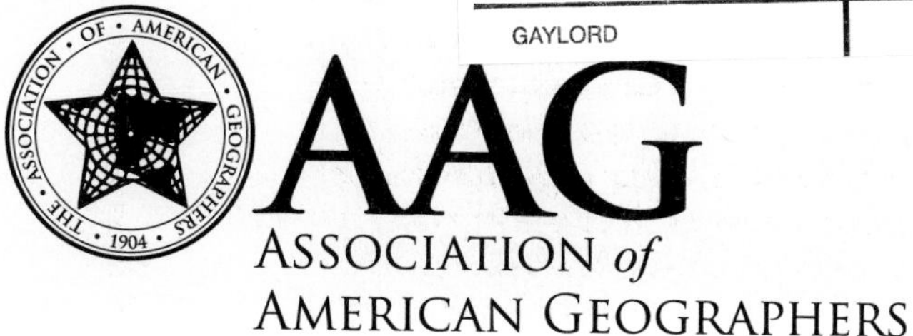

THE PEARSON AAG SERIES

Solem/Foote, *Teaching College Geography: A Practical Guide for Graduate Students and Early Career Faculty*

Solem/Foote/Monk, *Aspiring Academics: A Resource Book for Graduate Students and Early Career Faculty*

AAG Community-Based Website for New Scholars
www.pearsonhighered.com/aag

Solem/Foote/Monk, *Practicing Geography: Careers for Enhancing Society and the Environment*

Practicing Geography

EDITORS

Michael Solem
Association of American Geographers

Kenneth Foote
University of Colorado at Boulder

Janice Monk
University of Arizona

PEARSON

Boston Columbus Indianapolis New York San Francisco Upper Saddle River
Amsterdam Cape Town Dubai London Madrid Milan Munich Paris Montréal Toronto
Delhi Mexico City São Paulo Sydney Hong Kong Seoul Singapore Taipei Tokyo

Geography Editor: Christian Botting
Marketing Manager: Maureen McLaughlin
Project Editor: Anton Yakovlev
Editorial Assistant: Bethany Sexton
Marketing Assistant: Nicola Houston
Managing Editor, Geosciences and Chemistry: Gina M. Cheselka
Project Manager, Production: Wendy Perez
Full Service/Composition: PreMediaGlobal
Full Service Project Manager: Jenna Gray
Senior Art Specialist: Connie Long
Illustrator: Spatial Graphics
Cover Design: Seventeenth Street Studios
Photo Manager: Maya Melenchuk
Text Permissions Manager: Beth Wollar
Text Permissions Researcher: Warren Drabek
Operations Specialist: Michael Penne
Cover Photo Credits From top to bottom:
Yuri Arcursi/iStockphoto.com
Courtesy of Kevin McManigal
Greg Pickens/iStockphoto.com
Sprokop/Dreamstime.com

This material is based upon work supported by the National Science Foundation under Grant No. DRL-0910041. Any opinions, findings, and conclusions or recommendations expressed in this material are those of the author(s) and do not necessarily reflect the views of the National Science Foundation.

Library of Congress Cataloging-in-Publication Data

Practicing geography / Association of American Geographers; editors, Michael Solem, Kenneth Foote, Janice Monk.
p. cm. — (The Pearson AAG series in geography)
ISBN 978-0-321-81115-8
1. Geography—Vocational guidance. I. Solem, Michael. II. Foote, Kenneth. III. Monk, Janice J. IV. Association of American Geographers.
G65.P73 2013
910.23—dc23

2011043415

10 9 8 7 6 5 4 3 —EBM—15 14 13 12

PEARSON

ISBN-10: 0-321-81115-1; ISBN-13: 978-0-321-81115-8

CONTENTS

ABOUT OUR SUSTAINABILITY INITIATIVES

Pearson recognizes the environmental challenges facing this planet, as well as acknowledges our responsibility in making a difference. This book is carefully crafted to minimize environmental impact. The binding, cover, and paper come from facilities that minimize waste, energy consumption, and the use of harmful chemicals. Pearson closes the loop by recycling every out-of-date text returned to our warehouse.

Along with developing and exploring digital solutions to our market's needs, Pearson has a strong commitment to achieving carbon-neutrality. As of 2009, Pearson became the first carbon- and climate-neutral publishing company. Since then, Pearson remains strongly committed to measuring, reducing, and offsetting our carbon footprint.

The future holds great promise for reducing our impact on Earth's environment, and Pearson is proud to be leading the way. We strive to publish the best books with the most up-to-date and accurate content, and to do so in ways that minimize our impact on Earth. To learn more about our initiatives, please visit **www.pearson.com/responsibility**.

PREFACE

A World of Opportunity: Discovering and Managing Careers in Geography

Michael Solem, Kenneth Foote, and Janice Monk

“Itinerant” is one of many adjectives that aptly describe the work of Reena Patel, a Foreign Service Officer for the U.S. State Department. From her Peace Corps stint in Ghana to her dissertation field studies in Bangalore, Mumbai, and Ahmedabad, Reena credits her geography education and travel experiences for igniting her interest in international diplomacy. “It feels a little like being a professor,” she points out, “but instead of changing classes every semester, you change countries every few years.”

William Shubert never imagined that one day he’d spend the better part of a Friday on the National Mall in Washington, DC, dressed in a full-body zebra “morphsuit” to promote an upcoming “Great Migrations” TV special for his employer, the National Geographic Society (NGS). Only a few years earlier, NGS had declined his application for an internship. As William notes, his persistence paid off: “Not being good enough at the time doesn’t mean you’re not good enough ever.”

Carmen Tedesco’s interest in biology steered her toward the study of infectious diseases in college. Eventually, though, she realized that her interests extended beyond what she could view through a microscope. “Geography allowed a larger look at the issues related to disease,” she explains. Carmen now works for Development Alternatives, Inc., a Bethesda, Maryland-based development firm. “My work is really eclectic,” she says, “but that’s one of the main benefits of being a geographer.”

When Andrew Telfer’s environmental dream job proved elusive, he took the advice of his sister and pursued career opportunities in business and management. Years later, his business experience opened doors to a management position with Walmart Canada, where today he uses his geography knowledge to implement sustainability initiatives. “The stars aligned,” he recalls. “I had always wanted to get my career back to geography.”

The personal anecdotes you've just read come from just a few of the professional geographers whose work and experiences are shared throughout this book. As their experiences suggest, today is an exciting time to be a geographer. Whether you are just beginning as a major, taking classes toward a geographic information systems (GIS) certificate, working on an advanced degree, or considering a career change at a different point in your life, geography can lead to exceptional opportunities. Geographers are in demand for the range of knowledge, abilities, and skills they offer to businesses, government agencies, nonprofit organizations, and educational institutions. Few disciplines prepare individuals like geography does for thinking and working across the social and environmental sciences, in a host of technological fields, as well as in the arts and humanities (Murphy 2007). Geographers understand the practical and theoretical value of thinking spatially, whether as urban planners assessing the costs and benefits of a proposed highway, climatologists assessing the impacts of rising sea level, consultants advising a firm about moving into a new market, or human rights advocates working with refugees. These opportunities exist because geographic knowledge, skills, and perspectives help us understand and respond to environmental change, promote sustainability, recognize and cope with the rapid spatial reorganization of the economy and society, and leverage technological change to improve society and the environment (NRC 2010, 2–4).

This book is about finding and pursuing such career opportunities in geography. We have enlisted some of the most experienced and knowledgeable people in the discipline to provide information and advice we think will be useful in helping you define your career goals and decide how to attain them. Even if you are well along on your career plans, this book offers a rich mixture of ideas and personal examples that may suggest new or different career directions, now or in the future. In doing so, it differs from other career and job-hunting guides you may have read.

In *Practicing Geography*, we focus on the ways geographical knowledge, skills, and perspectives contribute to particular jobs and careers. This isn't to deny that many other professional skills and abilities are involved in career success. Indeed, you will often hear how much employers value so-called soft and people skills such as interpersonal communication, conflict resolution and negotiation, creative problem solving, strategic thinking, coaching and supervising, and being able to work effectively in teams and collaborative groups. As always, transferable skills such as writing and oral communication are broadly valued in all types of workplaces, as are research skills and skills at managing finances and budgets (Nerad et al. 2007; Babbit et al. 2008). Being adaptable to changing circumstances helps, too, as does showing an interest in learning about a company or agency's vision and culture—and if, like William Shubert, your employer asks you to dress like a zebra, just view it as a geographer's version of "casual Fridays."

At the same time, we think it is vital to articulate as forcefully as possible the value of geographical knowledge and abilities in the workforce across a wide range of careers. It is too easy, we think, to understate the many skills and perspectives geographers may bring to their work, such as spatial thinking; an interdisciplinary perspective; abilities in GIScience, cartography, and visualization; experience in field methods; a sense of the complex interactions between humans and environment; sensitivity to the distinctiveness of places; and a global perspective (Solem, Cheung, and Schlemper 2008, 367). The authors of the chapters in this book do an excellent job of articulating the value of these skills within many different employment sectors. In addition, they suggest strategies for getting the most out of classes, internships, study abroad programs, and other educational and professional opportunities that lead to good careers. Their message is that, by planning ahead, you can strengthen some of the skills likely to be valuable in your professional work, regardless of whether you are currently a student or have already moved into the workforce.

But as we know, life is unpredictable—full of curveballs and popups that upset even the most careful game plan. Sometimes we simply can't find our "dream job" because of an economic downturn or events beyond our control. Many of the professional geographers you'll meet in this book have dealt with similar challenges. By returning to school or pursuing a new career path seemingly unrelated to geography, they developed complementary skills that enabled them to practice their geography abilities in different, but equally rewarding ways. Many have found that their geographical perspective helped give them the flexibility to adapt to changing circumstances. We hope this book strengthens your ability to use the same perspective on your career as well as how to communicate your geographic abilities in a language that resonates with a wide range of potential employers.

The chapter authors have tried to strike a balance between information that will help you make good plans for your career and how to respond effectively to unexpected challenges and opportunities. Sometimes great opportunities appear unexpectedly, just as sudden life events may alter our plans—the birth of a child or the need to care for a partner or parent. Alternatively, change may come gradually as new passions and interests develop. It may seem contradictory, but having a well-defined career plan can actually help in times of uncertainty. The process of career planning can provide insights about your professional identity as a geographer: What kind of geographer am I? What has my academic preparation given me the skills to do? How do my abilities relate to what I care about? Where can I make a difference? Asking these questions in advance can help you move more quickly toward your career goals.

WHO PRACTICES GEOGRAPHY?

We hope to reach a very broad audience with this book, reflecting the full diversity of the discipline itself. Our aim is to cover the amazing variety of ways that geography is contributing to the workplace and to improving the quality of life in places and environments. As you read the chapters, you'll be introduced to geographers from all quarters of the discipline: human and cultural geographers, environmental and physical geographers, as well as those who specialize in geographic information technology and science.

The remarkable growth of GIS and other geospatial technologies in the workforce over the past two decades is certainly a driving force behind geography's renaissance as a science and profession, but we also wish to highlight the wider value of geographic practice. Although many of our profiles were selected to showcase the work of people who graduated with undergraduate or graduate degrees in geography, it's important to note that having a geography degree isn't a prerequisite for practicing geography. Indeed, many people are engaged in work that could be reasonably considered to be geographical in nature. Not everyone who works with geographic technologies, analyzes maps and satellite imagery, conducts fieldwork, or studies neighborhood demographics holds the job title of "geographer," which, after all, is an uncommon title even for people with geography degrees!

Professionals who routinely apply geography or who work with maps and other geographic technologies may hold positions as resource managers, civil engineers, landscape architects, environmental scientists, teachers, or in scores of other occupations. As preparation for their careers, some of them may have pursued a minor in geography or a certificate in GIS as a working professional. Others may have learned geographic techniques through an internship or a professional development workshop offered by their employer. So, rather than force people into arbitrary categories, we included all of these identities in our conception of a professional who practices geography.

For the same reason, we chose *Practicing Geography* as the title of this book. The term *practicing* is usually applied to professions such as medicine, law, and engineering, but we think it applies just as well to geography. It implies a close connection between theory and practice and acknowledges that both are intertwined at a very fundamental intellectual and scientific level. Not so long ago, considerable debate focused on distinguishing between pure and applied geography (Kenzer 1989a, 1989b). In hindsight, we wonder whether geography could ever be viewed in such sharp binary terms. Certainly there are differences in the career paths geographers follow, but not necessarily in the underlying scientific and intellectual challenges they face in their work. Many geographers are doing theoretically rich and highly creative work in businesses, government agencies, and nonprofits. They, for instance, draw on their geographic expertise and apply geographic principles when they devise transportation plans, make economic forecasts, predict the spread of a forest fire or disease outbreak, and advise policy makers on national security issues. Some geographers may apply for grants to fund a start-up company, while others mentor co-workers, lead tourists on an interpretive walk through a national park, or volunteer for community organizations. In other cases, instead of publishing the results of basic research in academic journals (though some do), they may write a brief report summarizing the key outcomes of their marketing research for a corporate board.

We've designed *Practicing Geography*, then, to help you recognize the continuum of practices and related career opportunities available to geographers in all areas of employment, while concentrating mainly on opportunities outside of the academy. Despite the fact that most geographers will ultimately enter a nonacademic private or public sector career, the literature has been strangely silent about their experiences and what can be done to better support them. Whereas some academic geography departments maintain close relationships with employers via internship programs and the use of alumni and professionals as adjunct instructors, other departments pay relatively little attention to workforce issues that can affect the employment prospects of their graduates. To change this situation, we think it is necessary to look more broadly at how we can improve professional development and career preparation in geography.

IMPROVING PROFESSIONAL DEVELOPMENT AND CAREER PREPARATION IN GEOGRAPHY

This book may be one step toward improvement, but more needs to be done to increase support for geographers moving into professional life, no matter their career path. In advocating this change, we aren't implying a need for overhauling graduate or undergraduate programs, but simply suggest ways to strengthen the discipline by increasing awareness of career opportunities and effective preparation strategies. For example, over the last decade we have tried to help develop a community of support for early career geographers moving into academic positions. This Geography Faculty Development Alliance project (Foote 2010; GFDA 2011) has focused on helping early-career faculty understand and find balance among their many teaching, research, and service responsibilities, as well as to consider realistically how their personal and professional lives intersect and interconnect. GFDA has involved residential workshops, panel and paper sessions at professional meetings, and the publication of *Aspiring Academics* (Solem, Foote, and Monk 2009) and *Teaching College Geography* (Solem and Foote 2009b; EDGE 2011).

In *Aspiring Academics* and *Teaching College Geography* we highlighted the value of better support for early-career faculty; here we focus on the benefits of improving support for geographers moving into public and private sector careers beyond the academy (Solem and Foote 2009a; EDGE 2011). This effort will involve work on several fronts. First, there is a need to situate geography broadly within the context and trends of the larger scientific, professional, and technical workforce. There has been rapid growth in career opportunities and need for geographic expertise across the public and private sectors, from small business in local communities to international companies and nongovernmental organizations (NGOs). At the same time, some of these opportunities are in new areas of employment for geographers, and often employers are seeking new combinations of geographical and other skills. The U.S. Department of Labor, for instance, projects "faster than average" or "much faster than average" growth in jobs for geographers, geoscientists, cartographers, urban and regional planners, and other geographic professionals, with projected needs of upwards of 15,000 additional employees in each of these career fields between 2008 and 2018 (U.S. Department of Labor 2010). Tracking these trends has two benefits: (1) it helps guide students to careers that fit their interests and abilities; and (2) it suggests ways in which educational programs can be improved and upgraded to help students gain the sorts of knowledge, abilities, and skills needed in emerging fields of the workforce. For more detailed statistical information about career and employment trends in geography, check out AAG's Jobs and Careers in Geography website (**www.aag.org/careers**) where you will find a database tracking categories, types, and numbers of geography-related occupations, as well as related salary trends and hiring outlooks.

Second, change will always involve continuing efforts to communicate geography's value to employers. Dialogue and engagement are needed to dispel misunderstandings about geography and what geographers do, as well as to make the case for what geography and geographers can do. Despite the present ubiquity of geotechnologies in everyday life, the use of maps and geovisualizations in media, and the rise of location-based services, there will always be a need for geographers to articulate clearly and argue persuasively for the knowledge and skills our discipline can contribute to business, government, and nonprofit organizations.

Third, improvements can be made in our undergraduate and graduate programs in the ways that educators inform and advise students about workforce trends and how to connect their skills and abilities to the needs of employers. For example, the AAG conducted surveys and interviews with employer organizations to identify some of the core skills required for work in different geography-related occupations (Solem, Cheung, and Schlemper 2008). This work found a broad need for professional geographers to be able to perform general skills ranging from time management and writing to working effectively with computers and applying critical thinking skills for problem solving. With regard to geographic skills, many employers seek individuals who possess competency in field methods, have the ability to work across disciplinary boundaries, and can apply principles of spatial thinking and other forms of geographical analysis to support a variety of tasks and needs.

Although we would argue that many of these general and geographical skills are included in typical geography curricula, it would seem that there is substantial potential for developing methods for cross-training students through greater use of pedagogies such as problem-based learning, inquiry, and service learning, as well as providing more students with opportunities to apply their abilities in professional settings through internships. In recent years, master's programs have, on some campuses, taken the lead in this area. This may be the result of the master's degree gradually becoming the "new" bachelor's degree the United States. It is now the entry-level credential for many geographical careers and has, as a consequence, become the focus of much innovation in terms of program structure, delivery, and new models of pedagogy. Within American master's programs, for example, there are now fifteen different types of degrees, such as geography and law; geography and environmental studies; geography and regional planning; and GIScience. And, increasingly, these programs are relying on some type of e-learning. The new Professional Science Master's (PSM) degree (Council of Graduate Schools 2011), which has been developed in a few geography programs, focuses explicitly on cross-training between discipline-specific skills and other professional skills such as business, finance and marketing, and legal and regulatory affairs. To strengthen academic–industry ties, PSM programs are overseen by a board representing different employer organizations, who recommend ways of incorporating professional training in written and oral communication, ethics, teamwork, and other aspects of organizational culture.

Improvements can also be made at the doctoral level. In many doctoral programs the hidden assumption is that all or most students are moving toward academic careers (Monk, Foote, and Schlemper in press). Although there is no question that our Ph.D. programs are vital to producing scholars capable of carrying out valuable basic research, the doctoral degree is also changing. It remains an apprenticeship in research, but has also become a qualification for a wide range of challenging careers and executive positions in business, government, and nonprofit organizations. Indeed, as Fagan and Wells (2004) have noted, students tend to report the greatest satisfaction with curricula that prepare them for a variety of careers and allow them to make informed choices among the alternatives. Unfortunately, as Wulff and Nerad (2006, 87) have noted:

> *Often the goals of students are very different from the goals and intentions with which they are being prepared. In some cases, the students may not be expressing those goals openly to advisors and mentors with whom they are interacting most closely, thus creating challenges for faculty/staff who work closely with students in the socialization process.*

This happens because:

> *. . . faculty members in some disciplines still do not accept that there is a widening range of career paths for contemporary students. Instead, those faculty continue to prepare their students as though all . . . are going to become faculty, while in reality only about half of all Ph.D. recipients ultimately do so.* (Wulff and Nerad 2006, 89)

These patterns are not unique to geography programs, as evidenced in similar workforce analyses in neighboring disciplines such as sociology (Spalter-Roth, Jacobs, and Scelza 2010) and the geosciences (Gonzalez and Keane 2011). In all of these cases, the issues ahead revolve around attitudes and perceptions as much as concrete reforms and innovation. Like earlier calls for reforming doctoral education by improving preparation in teaching, publishing, grantsmanship, institutional service, and other areas of academic practice, there are many ways of broadening the focus on careers while maintaining the traditional strengths of the Ph.D. as a research degree. Such changes are already underway in Ph.D. programs overseas. The Scottish government, for instance, is now funding studentships that train doctoral students to become employable in both academic and government settings through annual month-long residencies in public agencies, thereby facilitating knowledge transfer between academic and policymaking sectors (Reid and McCormick 2010). Many chapters in this book offer signposts for U.S. academic programs considering similar innovations.

It is also important to note that a lot of traditionally "academic" skills, such as research, public speaking, and teaching, are also applicable outside of the university setting. As Basalla and Debelius note (2007, 28): "The boundary between academia and the post-academic world is more permeable than you may realize." Students are often being equipped with important skills for the nonacademic workforce, but they may simply not be aware of how they can be applied within these contexts.

Finally, improved professional development and career preparation can help support the participation of women and of underrepresented ethnic and racial groups in geography. This issue has long been of concern within the discipline (Monk 2004; Monk et al. 2004; McEwen et al. 2008; Schlemper and Monk 2011; Monk 2012). Overall, white men are the most likely to desire faculty careers, followed by (in descending order) men of color, white women, and women of color (Golde and Dore 2001). Geography students across all demographic categories, but especially women and minorities, indicate a strong interest in pursuing careers for social and environmental change (Solem, Lee, and Schlemper 2009). Professional development can help enhance the discipline's diversity through recruitment and retention in part because often a mismatch exists between the implicit knowledge and skills needed to succeed in professional careers and the topics covered explicitly in graduate curricula, advising, and mentoring. Factors such as gender, sexuality, age, family status, nationality, race, and ethnicity often play a role in whether students are helped or hindered in mastering this "hidden" curriculum. By exposing these factors to scrutiny, more equitable access is possible.

HOW THIS BOOK IS ORGANIZED

We've sequenced the chapters in a way that reflects many of the concerns and issues of relevance to geographers at different stages of professional development, starting at the very beginning of their undergraduate studies proceeding onward through the job search and initial years of employment. The book concludes with a set of chapters examining overarching issues central to healthy, enriching, and successful careers over the lifespan. Throughout, you'll encounter advice and explicit cross references among the chapters to underscore the importance of having a developmental perspective of your career and being open to change and alternatives.

The first section (Chapters 1–5) explores what it takes to make a successful transition from college or graduate school and establish oneself in the early years of a career. Leading off the collection are two chapters that are in many ways foundational to all the others. These are Alyson Greiner and Thomas Wikle's chapter on key issues and strategies involved in career planning, and a chapter on geography education and career readiness by Joy Adams, Niem Huynh, Joseph Kerski, and Brent Hall. This second chapter is of particular interest because it focuses on the knowledge and skills geographers offer employers as well as how geographers can better communicate these and their own abilities to employers. In the third chapter, Joy Adams addresses "transitioning" and issues related to moving from one career field to another, an activity that is more common in today's workforce than it once was. Internships are recommended by many as essential to career preparation, a topic raised in the fourth chapter by Denise Blanchard, Mark Carter, Rob Kent, and Chris Badurek. Chapter 5 by Tina Cary highlights the value of professional networking from the start of a career onward.

The middle chapters (Chapters 6–13) examine the variety of career opportunities available to geographers in the public and private sectors. Here, you will learn some of the reasons why geographers are attracted to specific careers and some of the ways they are contributing with their geographic expertise and abilities. These chapters focus on particular areas of employment and were written by professionals with working experience in state and federal government, the corporate world, and nonprofits, and also those who have worked independently as a consultant. The authors discuss strategies for preparing for employment and make a point to highlight key skills and competencies that are important for work in a particular context. Chapter 6 by Bill Bass and Rich Quodomine examines careers in state and local government, followed by Chapter 7 on federal government careers by Allison Williams, Molly Brown, Erin Moriarty and John Wertman. Next up are three chapters exploring careers in for-profit and nonprofit organizations: Amy Blatt and Michael Ziolkowski's chapter on careers in corporations and large businesses, Lia Shimada and Jeremy Tasch's on nonprofits and NGOs, and Kelsey Brain's chapter on starting a small geography business. Many of the issues these authors raise are explored from different angles in the subsequent chapters, beginning with Carrie Mitchell and Mélanie Robertson's chapter on working internationally, followed by Susan Heffron writing on opportunities for geographers in education, and finally Angie Donelson and Susi Moser on consulting opportunities.

The book concludes with three chapters (Chapters 14–16) that focus on career issues of concern to professionals regardless of their occupation or career stage. In Chapter 14, Jan Monk addresses work and life relationships, but raises bigger questions about how we frame the terms *work* and *life* and how we see our personal and family lives intersecting with our careers. Recent years have seen an increasing mismatch between the demographics of the workforce and workplace policies and cultures. Dual-income families, single-parent families, single-income individuals living alone or with partners, unmarried couples with children, blended families, and other personal and family arrangements have become more common. At the same time, the workplace tends still to be geared toward the traditional nuclear family in which personal and family responsibilities can be shared by partners, whether to care for each other, their children, or aging or ill family members. Next, Francis Harvey's chapter on ethics introduces issues that are critical to our professional careers, but often not addressed explicitly in undergraduate and graduate curricula. He discusses a variety of resources on practical ethics for geographers, and how individuals can make a difference through ethical conduct. In the final chapter, Pauline Kneale and Larch Maxey discuss lifelong professional development and how we grow and develop professionally across all career stages. They stress the value of taking time to reflect on values and aspirations, to evaluate career plans and revisit them regularly, and to plan ahead to make the most of our opportunities, skills, experiences, and interests.

NEXT STEPS AND ACKNOWLEDGMENTS

We see this book as a starting place—for you in your career, but also for geography as a discipline. Geography is experiencing a renaissance, offering an exciting array of career opportunities across a wide range of fields. However, sustaining this momentum will involve further work along a number of fronts. This will include developing better dialogue between professional geographers and those leading our academic programs. Moving forward will also require us to assess and improve undergraduate and graduate programs on a continual basis, and to be proactive in supporting students rather than passively assuming they can fend for themselves. As a professional community we must stay aware of trends in science and the workplace, while thinking creatively about improving and reforming education, raising awareness, and understanding how geographical training can serve as a gateway to a wide range of careers. In the long run, geography will be judged by how it contributes to business, government, nonprofit organizations, to the education of the public, as well as to science and scholarship. This means building a stronger sense of community and common purpose among geographers and acting on the many challenges and opportunities that lie ahead.

In any undertaking such as this book, there are many individuals whose generosity and engaged, thoughtful advice made the project possible and such a pleasure to work on. The authors did a remarkable job of writing and revising their chapters on tight deadlines. We appreciate very much the time volunteered by the professionals who are profiled in this book, and we extend our deep thanks to Joy Adams and Mark Revell at the AAG for conducting the interviews. We also benefited from the advice of many reviewers, including Sonia Arbona, Vanessa Bauman, Sarah Bednarz, Kristina Bishop, Larry Carlson, Judy Colby-George, Jane Daniels, Serge Dedina, David DiBiase, Tom Dwyer, Kate Edwards, Pam Foti, John Frazier, Rob George, Kobena Hanson, Sandy Holland, Ann Johnson, Rob Kent, Phil Klein, Susi Moser, Lisa Nungesser, Ben Ofori-Amoah, Linda Peters, Richard Rhoda, Sue Roberts, Rich Schultz, Nate Sessoms, Fred Shelley, Isobel Stevenson, Trudy Suchan, George Thompson, Tom Wilbanks, and Nancy Wilkinson.

This book is one of the major outcomes of the AAG's Enhancing Departments and Graduate Education in Geography (EDGE) project. We would like to extend a special word of appreciation to Myles Boylan, our program officer at the National Science Foundation. NSF has supported the EDGE project's research, publication, and outreach activities since 2005, and we are grateful for the professional development opportunities and new collaborative relationships that the project has provided us and many others over the years. Through these investments, the AAG has been able to develop a large variety of programs at a national scale supporting students, early-career professionals, and the health of academic and training programs in geography.

Finally, we would like to thank Pearson Education, especially our managing editors Christian Botting and Anton Yakovlev who have supported our professional development publications for many years.

REFERENCES

Babbit, V., E. Rudd, E. Morrison, J. Picciano, and M. Nerad. 2008. Careers of geography PhDs: Findings from social science PhDs—Five+ years out. Seattle, WA: University of Washington, Center for Innovation and Research in Graduate Education. http://depts.washington.edu/cirgeweb/c/wp-content/uploads/2008/04/ss5-geography-report.pdf (accessed 31 May 2011).

Basalla, S., and M. Debelius. 2007. *"So what are you going to do with that?": Finding careers outside academia*. Revised edition. Chicago, IL: University of Chicago Press.

Council of Graduate Schools. 2011. Professional science masters. http://www.sciencemasters.com/ (accessed 31 May 2011).

EDGE (Enhancing Departments and Graduate Education in Geography), Association of American Geographers. 2011. http://www.aag.org/cs/edge (accessed 31 May 2011).

Fagan, A. P., and K. M. S. Wells. 2004. The 2000 national doctoral program survey. In *Strategies for enriching the preparation of future faculty*, eds. D. E. Wulff, A. E. Austin, and associates, 74–91. San Francisco, CA: Jossey Bass.

Foote, K. E. 2010. Creating a community of support for early career academics. *Journal of Geography in Higher Education* 34 (1): 7–19.

GFDA (Geography Faculty Development Alliance), University of Colorado at Boulder, Department of Geography. 2011. http://www.colorado.edu/geography/gfda/gfda.html (accessed 31 May 2011).

Golde, C. M., and T. M. Dore. 2001. *At cross purposes: What the experiences of doctoral students reveal about doctoral education.* Philadelphia, PA: A report prepared for The Pew Charitable Trusts. http://www.phd-survey.org.

Gonzalez, L., and C. Keane. 2011. *Status of the geoscience workforce.* Washington, DC: American Geological Institute.

Kenzer, M. S., ed. 1989a. *Applied geography: Issues, questions and concerns.* Dordrecht, NL: Kluwer.

Kenzer, M. S., ed. 1989b. *On becoming a professional geographer.* Columbus, OH: Merrill Publishing.

McEwen, L., J. J. Monk, I. Hay, P. Kneale, and H. King. 2008. Strength in diversity: enhancing post-graduate education. *Journal of Geography in Higher Education* 32 (1): 101–119.

Monk, J. 2004. Women, gender, and the histories of American geography. *Annals of the Association of American Geographers* 94: 1–22.

Monk, J. 2012. Women and minorities in geography. In *21st century geography: A reference handbook*, ed. J. P. Stoltman, 379–386. Thousand Oaks, CA: Sage.

Monk, J., J. Droogleever Fortuijn, and C. Raleigh. 2004. The representation of women in academic geography: Contexts, climate and curricula. *Journal of Geography in Higher Education* 28: 83–90.

Monk, J. J., K. E. Foote, and M. B. Schlemper. In press. Graduate education in U.S. geography: Students' career aspirations and faculty perspectives. *Annals of the Association of American Geographers.*

Murphy, A. B. 2007. Geography's place in higher education in the United States. *Journal of Geography in Higher Education* 31: 121–141.

NRC (National Research Council). 2010. *Understanding the changing planet: Strategic directions for the geographical sciences.* Washington, DC: National Academies Press.

Nerad, M., E. Rudd, E. Morrison, and J. Picciano. 2007. Social science PhDs—Five+ years out: A national survey of PhDs in six fields—Highlights report. Seattle, WA: University of Washington, Center for Innovation and Research on Graduate Education.

Reid, L. A., and McCormick, A. 2010. Knowledge transfer at the research–policy interface: The geography postgraduates' experiences studentships. *Journal of Geography in Higher Education* 34 (4): 529 – 539.

Schlemper, M. B., and J. J. Monk. 2011. Discourses on "diversity": Perspectives from graduate programs in geography in the United State. *Journal of Geography in Higher Education* 35 (1): 23–46.

Solem, M. N., I. Cheung, and M. B. Schlemper. 2008. Skills in professional geography: An assessment of workforce needs and expectations. *Professional Geographer* 60: 356–373.

Solem, M. N., and K. E. Foote. 2009a. Enhancing departments and graduate education in geography: A disciplinary project in professional development. *International Journal of Researcher Development* 1 (1): 11–28.

Solem, M. N., and K. E. Foote, eds. 2009b. *Teaching college geography: A practical guide for graduate students and early career faculty.* Upper Saddle River, NJ: Prentice-Hall.

Solem, M., J. Lee, and M. B. Schlemper. 2009. Departmental climate and student experiences in geography graduate programs. *Research in Higher Education* 50 (3): 268–292.

Solem, M. N., K. E. Foote, and J. J. Monk, eds. 2009. *Aspiring academics: A resource book for graduate students and early career faculty.* Upper Saddle River, NJ: Prentice-Hall.

Spalter-Roth, R., J. Jacobs, and J. Scelza. 2010. *Still a down market: Findings from the 2009/2010 job bank survey.* Washington, DC: American Sociological Association.

U.S. Department of Labor. 2010. O*NET Online. http://online.onetcenter.org (last accessed: 9 June 2011).

Wulff, D. H., and M. Nerad. 2006. Using an assessment model as a framework in the assessment of doctoral programs. In *The assessment of doctoral education: Emerging criteria and new models for improving outcomes*, eds. P. L. Maki and N. A. Borkowski, 83–108. Sterling, VA: Stylus.

Profiles of Professional Geographers

Throughout the chapters in this book, you will find profiles of professional geographers working in various positions for business, government, and nonprofit (BGN) organizations. The profiles are based on interviews conducted by Dr. Joy Adams and Mark Revell at the AAG, with three added by Jan Monk for her chapter.

Given the many professional careers available to geographers, it was a challenge to select which ones to spotlight in the book. When we began this project, we wanted to ensure that we acknowledged the contributions being made by human geographers, physical and environmental geographers, as well as those who specialize in geographic information science and geospatial technology. We also strove to represent the work, interests, and experiences of early career and senior professionals, men and women, and individuals of diverse cultural backgrounds working in different regions of the U.S. and abroad.

The profiles offer real-world examples of how geographers practice their craft in different industries and sectors of employment. They also touch on some of the broader themes affecting the professional and personal lives of geographers. Although you may not find an example of the specific career that interests you, we think the available profiles offer lessons and advice that can benefit nearly anyone seeking a career in BGN organizations.

Individual profiles were matched with chapters based on their relevance to an overriding topic or theme; however, as you read the profiles you will discover many overlapping issues and complementary ideas with other chapters in the book. In addition, we will continue to publish new profiles in the AAG Newsletter (**www.aag.org/cs/publications**) and on the AAG Jobs and Careers website (**www.aag.org/careers**).

We hope that you find the profiles to be inspiring and insightful resources that further illustrate the ideas and topics presented in the chapters.

—Michael Solem, Ken Foote, and Janice Monk

Name	Title	Employer	Profile location
Arbona, Sonia	Medical Geographer	Texas Department of Health	Chapter 3
Baber, Max	Director of Academic Programs	U.S. Geospatial Intelligence Foundation	AAG Jobs and Careers website
Barnes, Mark	Eagleton Governor's Executive Fellow	New York-New Jersey Port Authority	Chapter 4
Boniche, Armando	Circulation Director	Miami Herald	Chapter 8
Brinegar, Sarah	Social Science Analyst	U.S. Dept. of Justice	Chapter 7
Carney, Kristen	Co-founder	Cubit Planning	Chapter 10
Daniels, Jane	Director of Preservation Programs	Colorado Preservation	Chapter 9
Dedina, Serge	Executive Director	WildCoast	Chapter 9
Dwyer, Tom	Principal	Dutch Hill Consulting	Chapter 15
Fearn, Steve	Director, Real Estate Market Research	Ross Stores, Inc.	Chapter 2
Finkbeiner, Karl	Operations Analyst	Ingram Micro	Chapter 8
Gettig, Tara	Environmental Education Specialist	Pennsylvania State Parks	Chapter 12

(Continued)

Name	Title	Employer	Profile location
Haug, Dan	IT Director	Confederated Tribes of the Umatilla Indian Reservation	Chapter 6
Henderson, Marsha	Vice President, External Affairs	SUNY-Buffalo	AAG Jobs and Careers website
Higgins, Jim	Regional Manager	Esri	AAG Jobs and Careers website
Huynh, Niem	Senior Researcher	Association of American Geographers	AAG Jobs and Careers website
Kooy, Michelle	Research Fellow, Water Policy Programme	Overseas Development Institute	Chapter 11
Letham, Glenn	Editor	GISuser.com	Chapter 5
Lewis, Nancy Davis	Director of Research Programs	East-West Center	Chapter 16
Lyon, Elizabeth	Research Geographer	U.S. Army Corps of Engineers	AAG Jobs and Careers website
Masó, Carmen	GIS Analyst	U.S. Environmental Protection Agency	Chapter 14
Messen, Dmitry	Manager, Socio-Economic Modeling Program	Houston-Galveston Area Council	Chapter 6
Miglarese, Anne Hale	Principal Consultant	Booz Allen Hamilton	Chapter 3
Morales, Anthony	Co-founder	Cubit Planning	Chapter 10
Moye, Valerie	Green Infrastructure Coordinator	City of Chattanooga	Chapter 1
Ofori, Esther	Analyst, Marketing Statistics	Macy's	Chapter 8
Patel, Reena	Public Diplomacy Officer	U.S. Department of State	Chapter 11
Pearson, Kate	Strategic Partnerships Manager	Habitat for Humanity International	Chapter 9
Prinsen, Scott	Meteorologist	KTBC Fox 7 News	AAG Jobs and Careers website
Ratcliffe, Mike	Assistant Division Chief	U.S. Census Bureau	Chapter 7
Selkowitz, Dave	Research Geographer	U.S. Geological Survey	Chapter 7
Shubert, William	International Editions Coordinator	National Geographic	Chapter 4
Tedesco, Carmen	Senior Spatial Planning and Development Specialist	Development Alternatives, Inc.	Chapter 1
Telfer, Andrew	Manager of Sustainability	Walmart Canada	Chapter 8
Tian, Nancy	Geographer	U.S. Environmental Protection Agency	AAG Jobs and Careers website
Trevino, Paula Ann	Grade-Level Principal	Helix Charter High School	Chapter 12
Turnbull, Ken	Accredited Land Consultant	Self	Chapter 5
Votteler, Todd	Executive Manager of Governmental Relations and Policy	Guadalupe-Blanco River Authority	AAG Jobs and Careers website
Wadwani, Ashok	Founder and President	Applied Field Data Systems	Chapter 10
Welborn, Jason	Biological Science Technician	National Park Service	Chapter 2
Young, Emily	Senior Director, Environment Analysis & Strategy	The San Diego Foundation	Chapter 9
Young, Jeff	Business Development Manager	LizardTech	Chapter 14

ABOUT THE EDITORS

Michael Solem joined the staff of the Association of American Geographers (AAG) in 2003 as Educational Affairs Director. At the AAG, he directs many federally funded initiatives including the Enhancing Departments and Graduate Education in Geography (EDGE) and Center for Global Geography Education (CGGE) projects, both funded by the National Science Foundation (NSF). Dr. Solem serves as the North American coordinator of the International Network for Learning and Teaching Geography in Higher Education (INLT), is associate director of the Grosvenor Center for Geographic Education at Texas State University–San Marcos, and is Treasurer for the International Geographical Union's Commission on Geographical Education. He twice received the *Journal of Geography in Higher Education's* award for promoting excellence in teaching and learning for his research on faculty development and graduate education in geography.

Kenneth Foote is Professor of Geography and former Department Chair at the University of Colorado at Boulder. Much of his research focuses on landscape history, GIScience, cartography, and geography education, but over the past decade his work has concentrated on improving support for early career geographers and strengthening leadership training in geography. He has led the NSF-funded Geography Faculty Development Alliance since 2002 and is co-PI of the Enhancing Departments and Graduate Education in Geography project with Michael Solem and Janice Monk. Dr. Foote has served as president of the National Council for Geographic Education (2006) and president of the Association of American Geographers (2010–2011). He received the Association of American Geographers' 1998 J. B. Jackson Prize for his book *Shadowed Ground: America's Landscapes of Violence and Tragedy* and the association's 2005 Gilbert Grosvenor Honors in Geographic Education.

Janice Monk is Research Professor in the School of Geography and Development at the University of Arizona where she coordinates a graduate course in professional development. She previously served for twenty-five years as Executive Director of the Southwest Institute for Research on Women, leading research, educational, and community outreach projects in the southwestern United States and northwestern Mexico. Dr. Monk is also Adjunct Professor of Human Geography at Macquarie University (Australia). She has long been involved with research and projects in higher education and in gender studies and has served on various editorial boards and advisory panels. Dr. Monk is active in the International Geographical Union's Commission on Gender and Geography and is the long-standing editor of its newsletter. A former President of the Association of American Geographers, she has received several awards, including Lifetime Achievement Honors of the AAG and the Australia-International Medal, Institute of Australian Geographers. She is Co-Principal Investigator of the AAG's Enhancing Departments and Graduate Education in Geography project.

ABOUT THE AUTHORS

Joy K. Adams is a Senior Researcher at the headquarters of the Association of American Geographers in Washington, DC, where she contributes to projects on professional development and careers for geographers, geographic education, and diversity within the discipline. During her career, Dr. Adams has worked in each of the BGN (business, government, and nonprofit) sectors, in addition to having held faculty positions at Humboldt State University, Texas State University-San Marcos, and the University of Wisconsin-Oshkosh. Her research and teaching have focused on the social construction of ethnic and racial identities in the United States, cultural landscapes of North America, heritage tourism, qualitative methods, experiential learning, and research and writing instruction.

Christopher A. Badurek is an Assistant Professor of Geography and Planning at Appalachian State University specializing in GIScience, environmental planning, and natural hazards. Over the past ten years, he has worked on or managed GIS, database, and information technology projects for NSF, NASA, NGA, USAID, the National Park Service, the Multidisciplinary Center for Earthquake Engineering (MCEER) Information Service, and the North Carolina State Energy Office. Dr. Badurek currently serves as the Chair of the Environmental Perception and Behavioral Geography Specialty Group of the AAG, Secretary-Treasurer of the North Carolina Geographical Society, and Board Member of the North Carolina ArcGIS Users Group. He has supervised over seventy GIS internships in the past five years at a variety of private and government agencies.

William (Bill) Bass is Chief GIS Specialist in the Community and Environmental Planning Department of the Houston-Galveston Area Council (H-GAC), a regional planning agency for Greater Houston. In this role, Bill manages all aspects of GIS operations in the department which includes: GIS applications architecture and web-based GIS development, regional geospatial database development, and spatial analysis and cartography in support of socio-economic and environmental modeling. Bill holds a Master's of Science degree in Geography from Texas State University-San Marcos, and is a certified GIS professional (GISP). His professional interests include geospatial applications development, spatial data infrastructures, use of GIS in hazards research, cartographic visualization, and Web-based mapping systems.

Denise Blanchard is Professor of Geography in the Department of Geography at Texas State University-San Marcos. She specializes in natural and environmental hazards, communication geography, historical geography, environmental economics, and research methods. Dr. Blanchard has been involved in scholarly research and teaching for more than twenty-three years at Texas State University as well as various departmental administrative positions, including four years as Internship Director and eight years on the Internship Advisory Committee. During that time, she expanded the internship program and increased the number of geography interns from approximately 30 per year to more than 110 annually, with a substantial number of internships leading directly to employment. She has been recognized for outstanding teaching and mentoring with several awards and nominations, including the National Council for Geographic Education's Distinguished Teaching Award.

Amy J. Blatt is a Senior Health Informatics Analyst in the Informatics and Analytics group at Quest Diagnostics. Her main responsibilities include producing the clinically relevant Quest Diagnostics HealthTrends™ reports as well as supporting new marketing and sales initiatives through customer-driven insights and analytics. Dr. Blatt has also held professional positions as Geodemographer at Marketing Systems Group and as Statistical Data Analyst at the Hartford Customer Services Group. She also had a faculty appointment at West Chester University of Pennsylvania. She currently serves as the guest editor of a special theme volume on "Geographic Opportunities in Medicine" in the *Journal of Map and Geography Libraries* and as the president of the Esri Mid-Atlantic Users Group.

Kelsey Brain has enjoyed blending her experience in geography and business throughout her career. With an MS degree in Geography and a BS degree in Business Management with a focus on Entrepreneurship and Marketing, she currently teaches geography as an Adjunct Instructor at Portland State University and works in Human Resources, advising the university on recruitment and retention strategies for employees. In addition, she has recently founded a side business selling personal solar devices via drop-shipment methods. Her most recent research examines a cross section of economics and geography by looking at transnational commodity flows, specifically the facilitation of Peruvian food imports to San Francisco.

Molly Brown is a research scientist at NASA Goddard Space Flight Center in Greenbelt, Maryland, and has a master's and doctorate in Geography from the University of Maryland, College Park. Dr. Brown has worked for over ten years with colleagues at Goddard to develop and use a long-term climate data record of vegetation based on satellite remote sensing information. Using these satellite data to investigate questions of climate, agricultural production, and human–environmental interaction has been the main focus of her work. Currently she is involved in exploring decision making and the use of remote sensing data in governmental and nongovernmental organizations whose work is influenced by climate and weather. Dr. Brown has received the Robert H. Goddard Award for Exceptional Achievement in Science in 2008 and the NOAA David Johnson Award in 2010.

Mark Carter is a Senior Lecturer at Texas State University-San Marcos where he has served since 2000 as Internship Coordinator in the Department of Geography. In this capacity, he has helped more than 700 undergraduate and graduate students prepare for, identify, and complete a career-related internship as part of their academic program. Mark provides Texas State University Geography students general career advising as well as specific instructions on assembling a career-planning toolkit, and helps current students connect with a network of geography alumni and other potential internship sponsors through traditional hallway job postings as well as an online job/internship website. Mark works closely with the Texas State University Career Services Department to coordinate and promote a variety of career-information and recruiting events of interest to geography students throughout the academic year on the campus of Texas State.

Tina Cary, founder and President of Cary and Associates, is a past President and a Fellow of the American Society for Photogrammetry and Remote Sensing (ASPRS). As a marketing consultant and keynote speaker, Dr. Cary has worked with senior executives around the world in all sectors of the mapping sciences from Fortune 100 companies to universities, federal agencies, and nonprofits. Before founding Cary and Associates, she was employed at Space Imaging, EOSAT, Rutgers

University, NASA Goddard Institute for Space Studies, and Purdue University. She considers her lifelong networking habits a key factor in her career success. Dr. Cary is an enthusiastic user of social media and has given presentations on the use of social media in the geospatial technology industry.

Angela Donelson, as president of Donelson Consulting, LLC, has conducted research, planning, implementation, and evaluation of community development strategies for nonprofit organizations, foundations, and state and local governments. Prior to incorporating her firm in 2006, she worked as the United States Department of Housing and Urban Development's (HUD) representative to Arizona's colonias, assisting dozens of underprivileged, rural U.S.-Mexico border communities and nonprofits. Dr. Donelson has also worked as a city and regional planner in the private and public sectors in southern Arizona, New Jersey, and Kansas. She holds a doctorate in Geography from the University of Arizona and is certified by the American Planning Association as a professional planner (AICP). Dr. Donelson has co-authored two books on U.S.-Mexico border community development with the University of Arizona Press.

Alyson L. Greiner is Associate Professor of Geography at Oklahoma State University. Her research interests include Australian society and identity, geographic education, landscapes of the New Deal, and cultural resource management. Dr. Greiner has served as principal investigator on numerous historic preservation research grants, and recently worked with several colleagues in her department to create a professional development seminar for geography graduate students. Dr. Greiner is the author of a college-level textbook, *Visualizing Human Geography: At Home in a Diverse World*. She has also helped develop interactive media about careers in geography. Dr. Greiner currently serves as the editor of the *Journal of Cultural Geography* and is a Regional Councillor of the Association of American Geographers. She has received the Distinguished Teaching Award from the National Council for Geographic Education.

G. Brent Hall is Dean and Head of the National School of Surveying at the University of Otago, New Zealand. He is also an Adjunct Professor in the School of Planning, University of Waterloo, Canada. Professor Hall has held the Belle van Zuylen Visiting Chair in Geography at the University of Utrecht (The Netherlands) (1992), an Erskine Fellowship at the University of Canterbury, New Zealand (2004), and a CIES Fellowship in Lima, Peru (2005). He has co-authored one book, edited another, written numerous book chapters, and published over sixty sole- or joint-authored papers in peer-reviewed international journals. Professor Hall received the Horwood Critique Prize from the Urban and Regional Information Systems Association (1997), a university-wide award in teaching excellence at the University of Waterloo (2004), and a national award for excellence in teaching Geography from the Canadian Association of Geographers (2006).

Francis Harvey is an Associate Professor at the University of Minnesota. His research interests include spatial data infrastructures, geographic information and sharing, semantic interoperability, and critical GIS. Dr. Harvey serves on the editorial boards of the *International Journal for Geographical Information System, Cartographica, GeoJournal*, and the *URISA Journal*. He published *A GIS Primer* with Guilford Press in 2008. His recent work includes projects in Poland considering discrepancies between the cadastre and land use, examining Return-on-Investment of parcel data in regional data sharing (MetroGIS), and developing a model curriculum for GIS ethics teaching.

Susan M. Heffron is Senior Project Manager for Geography Education at the Association of American Geographers. She works on geography education projects focusing on K-12 learners and teachers as well as preservice undergraduate education. She participated in the effort to produce the *Geography for Life, 2nd Edition* project for the National Geography Standards. A former high school teacher, she received a Distinguished Teaching Award from the National Council for Geographic Education. She has also worked on the instructional design and implementation of multiple online teaching resources. Dr. Heffron serves on the Geography Education National Implementation Project coordinating committee, the editorial board of *The Geography Teacher*, as well as numerous other projects related to the improvement of geography education.

Niem Tu Huynh is a Senior Researcher at the Association of American Geographers (AAG). She is working on a two-year project, "Establishing a Road Map for Large-Scale Improvement of K-12 Education in the Geographical Sciences," led by the National Geographic Society (NGS) and funded by the National Science Foundation. Prior to joining the AAG, Dr. Huynh was an Assistant Professor in the Department of Geography at Texas State University-San Marcos. In this position, she taught undergraduate courses and graduate classes focused on geographic education and technology. In 2011, Dr. Huynh won the Face-to-Face Category for the Teaching with Sakai Innovation Award.

Robert B. Kent is Department Chair and the James H. Ring Professor of Urban Studies and Planning at California State University, Northridge. Dr. Kent is also the department's internship coordinator. He holds a BA and MA from the University of California, Davis, and a Ph.D. from Syracuse University, all in geography. Between 2000 and 2008 he served as the department chair and graduate internship coordinator in the Department of Geography and Planning at the University of Akron. During that time, Dr. Kent supervised over 350 internships and wrote over $1.1 million in internship contracts (almost all of which represented graduate assistant salary funds) with local governments, nonprofits, and businesses. His scholarly interests include urban and regional planning, Latin America, cartography/GIS, and human geography.

Joseph Kerski is a geographer by training who believes that spatial analysis with GIS technology can transform education and society through better decision making using the geographic perspective. He has served as geographer and cartographer at NOAA, the U.S. Census Bureau, and the U.S. Geological Survey, and has taught online and face-to-face courses at primary and secondary schools, in community colleges, and in universities. Dr. Kerski serves as Education Manager for Esri in Denver, Colorado, focusing on GIS-based curriculum development, research in the implementation and effectiveness of GIS in education, teaching professional development institutes for educators, and fostering partnerships and communication to promote and support GIS in formal and informal education at all levels, both in the United States and internationally.

Pauline Kneale is Pro Vice Chancellor of Teaching and Learning at the University of Plymouth, UK, and Director of the Higher Education Academy Subject Centre for Geography, Earth, and Environmental Sciences. Professor Kneale was Professor of Applied Hydrology with Learning and Teaching at the University of Leeds until 2009. She is a National Teaching Fellow and holder of the Royal Geographical Society Taylor and Francis Award. Her recent research has focused on student skills, enterprise, and entrepreneurship in the higher education curriculum, and enhancing the role of teaching in higher education.

Larch Maxey is a Research Fellow with Plymouth University and Honorary Research Fellow with Swansea University, specializing in sustainable education. He has twenty years' experience of sustainability teaching, research, and practice, with over forty academic and popular publications, including a co-edited book on Low Impact Development (**http://lowimpactdevelopment.wordpress.com**).

Dr. Maxey has been active in his own lifelong professional development since he was seventeen and has founded, co-founded, or led over twenty distinct projects and enterprises, including two Royal Geographical Society Research Groups, Swansea University's Sustainability Forum, and Lammas (**http://www.lammas.org.uk**), a pioneering network that built the UK's first Low Impact hamlet of nine small holdings.

Carrie Mitchell is a Senior Program Officer at the International Development Research Centre (IDRC) in Ottawa, Canada. She manages research on climate change and urban environmental management in Southeast Asia, South Asia, and the Middle East. Before joining IDRC, Dr. Mitchell conducted research on urban change and informal waste recyclers in Hanoi, Vietnam and worked on an urban waste management project in Vientiane, Laos. She obtained her Ph.D. in environmental geography from the University of Toronto, and also holds a Master of Science in Urban Planning and a bachelor's degree in international development. Dr. Mitchell has published and lectured on informality, urbanization, waste management, and qualitative and quantitative research methods.

Erin Moriarty is currently living and working in Melbourne, Australia. As a certified business continuity professional, she is providing business continuity, emergency management, and risk management policy and program support to the public sector. Before moving abroad, Erin worked for the Ontario provincial government in Canada, coordinating successful emergency management, safety, and security programs. Erin has also worked extensively in the community development field at the local and municipal level, developing and implementing policies and programs in at-risk communities.

Susanne Moser is Director and Principal Researcher of Susanne Moser Research & Consulting, in Santa Cruz, California. She also is a Social Science Research Fellow at Stanford's Woods Institute for the Environment and a Research Associate of the Institute for Marine Sciences at the University of California-Santa Cruz. In her current research and work with local, state, and federal government agencies and nongovernmental organizations, Dr. Moser focuses on adaptation to climate change, especially in coastal areas, resilience, decision support, and effective climate change communication in support of social change. She contributed to the Fourth Assessment of the IPCC and has been selected as a Review Editor for the IPCC Special Report on "Managing the Risks of Extreme Events and Disasters to Advance Climate Change Adaptation" and as a lead author in the Fifth Assessment. Dr. Moser is a fellow of the Aldo Leopold Leadership, Kavli Frontiers of Science, Donella Meadows Leadership, and Google Science Communication Programs.

Richard D. Quodomine is a Transportation Analyst and Geographic Information Systems Coordinator at the New York State Department of Transportation's Public Transportation Bureau. He is responsible for grant administration and oversight of the state's massive public transit system, in addition to providing geographic, fiscal, and contract analysis for the bureau. He is the Co-Chair of the Environmental Justice Task Force's Mapping Workgroup and Parliamentarian for the NYS GIS Association. Mr. Quodomine has appeared throughout the nation as a speaker for career development in Geography and geographic education at colleges and public schools. He holds bachelor's and master's degrees from SUNY Buffalo's Geography Department.

Mark Revell is a Research Assistant at the headquarters of the Association of American Geographers in Washington, DC. A former AAG intern, he joined the staff full-time in

September 2010 and has since contributed to projects focusing on professional development and geography education. Mr. Revell is a graduate of George Washington University, where he received a Master of Arts degree in Geography in May 2010. His master's research focused on urban geography, population geography, and the geography of public space. He worked as a Crew Leader on the 2010 Census and interned with a regional planning commission in Virginia before entering graduate school. He currently resides in Arlington, Virginia.

Mélanie Robertson is a Senior Program Officer at the International Development Research Centre (IDRC) in Ottawa, Canada. She manages research projects on sustainable natural resource management in West Africa and Southern Africa, most of which focus on the use of geographic information systems in decision making. Before joining IDRC, Dr. Robertson worked at the Institut national de la recherche scientifique of the Université du Québec. There she coordinated international research projects on the political and environmental effects of modern urban transformations. Dr. Robertson has also worked in Indonesia, Vietnam, and China on urban change and the use of geographic information systems to inform decision making. She obtained her Ph.D. in political geography from the Université de Montréal and has also done postdoctoral research in geography at the Institut Français d'urbanisme of the Université de Paris VIII.

Lia Dong Shimada was born in Seattle, Washington, and studied English literature and environmental sciences at Wellesley College. In 2000, she was awarded a Thomas J. Watson Fellowship to work with reforestation movements in Nepal, Madagascar, and Ireland. From 2007 to 2010, Dr. Shimada served as Project Development Officer for Groundwork Northern Ireland, facilitating dialogue about racism, hate crime prevention, and conflict transformation in paramilitary-controlled communities. She is a trained mediator and completed her doctorate in geography at University College London. Dr. Shimada currently serves as Learning and Development Officer for the Methodist Church in Britain, and also volunteers as a mediator for a community-based nonprofit called Common Ground, where she is collaborating on a restorative justice initiative in East London.

Jeremy Tasch is Towson University's first interdisciplinary hire in Eurasian and global studies. He came to Towson from his previous position as an Assistant Professor at the University of Alaska, Anchorage, where he helped create the university's first geography and environmental studies department and first undergraduate degree program in international studies. Dr. Tasch is a principal investigator on a five-country, NSF-funded study of the geopolitics of climate change in the Arctic. A recipient of two Fulbright awards to the Russian Far East and the Kyrgyz Republic, he spent almost five years as country director of an international educational NGO in Azerbaijan, where he helped create the first multi-institutional educational center and library funded by the U.S. Department of State.

John Wertman serves as Senior Program Manager for Government Relations at the Association of American Geographers. His career has focused on positions in and around politics, and he previously served as Special Assistant to the Director of Presidential Letters and Messages at the White House during the Administration of President Bill Clinton. Mr. Wertman also held government relations positions at the Consortium of Social Science Associations, the University of Virginia Medical Center, and the American Urological Association. He interned on Capitol Hill for both former Congressman Tom Davis and former House Minority Leader Dick Gephardt. Mr. Wertman holds a BA in Government from the University of Virginia.

Thomas Wikle is Professor of Geography and Associate Dean in the College of Arts and Sciences at Oklahoma State University. He has served as principal investigator for several education projects supported by the National Science Foundation, including the Rural Alliance for Improving Science Education (RAISE). Dr. Wikle's published work on geographic information science and higher education has appeared in the *International Journal of Geographical Information Science, Transactions in GIS,* and *Computers, Environmental and Urban Systems.* His recent research has examined the distribution of wireless communications systems.

Allison Williams is a social geographer specializing in research addressing healthcare services, quality of life, informal caregiving, critical policy/program evaluation, and therapeutic landscapes. She currently is an Associate Professor at McMaster University (Hamilton, Ontario) and has held previous academic appointments at the University of Saskatchewan (Saskatoon, Saskatchewan) and Brock University (St. Catherines, Ontario). Dr. Williams is the recipient of a number of awards, including the Canadian Institutes for Health Research (CIHR)—Ontario Women's Health Council, Institute for Gender and Health Mid-Career Scientist Salary Award (2008–2013), the CIHR New Investigator Salary Award (2001–2006), and the Canadian Association of Geographers Julian M. Szeircz Prize (2003). She is currently leading a number of CIHR-funded programs of research.

Michael Ziolkowski is Assistant Professor of International Business at the College at Brockport. He earned his doctorate in economic geography from the State University of New York at Buffalo where he researched university and academic linkages. In 1994 and 1995, Dr. Ziolkowski attended TSM Business School at the Universitiet Twente, Enschede, Netherlands, as a fellow studying innovation and business development. From 1995 to 2009, he worked in many capacities for the Global Corporate Operations Department of the technology distributor Ingram Micro, Inc. Dr. Ziolkowski's research and teaching interests are in the area of Canada-U.S. trade, international business, social and economic development, college extension programs, supply chain management, and international education.

1

Part Strategy and Serendipity: A Candid Guide to Career Planning for Geographers

Alyson L. Greiner and Thomas A. Wikle

The word *career* derives from the French *carrière*, meaning "the ground over which a race is run" (OED Online 2011), but it may be more useful to think of a career as a series of transitions that define or punctuate different moments in our lives. By convention, we highlight our occupational trajectory, our skill sets, and the general contours of our work experience in the form of a résumé or curriculum vitae (CV) for prospective employers. On paper—or on the Web!—our careers may appear to be linear and self-directed, but if we probe a little deeper and ask questions about how a person obtained a particular job it is not unusual to learn about a chance meeting that turned into a new opportunity. Career planning, therefore, entails a mix of strategic planning to enhance our employability and self-awareness to create spaces for those unexpected developments that may lead to new or alternative job prospects.

In this chapter, we adopt an expansive view of career planning that sees it as a combination of education, training, networking, and other experiences contributing to professional development and career preparation. We also note that career planning is a process affected by a number of personal and situational contingencies related to family obligations, finances, the health of the job market, and many other factors. Drawing from our own experiences as well as a robust body of research (see, for example, Osipow and Fitzgerald 1996; Patton and McMahon 2006), we feel that career planning is an ongoing and dynamic process that each of us continuously participates in, sometimes more actively or purposefully than others. Over the course of our lives, we engage in career planning when we draw on feedback from others about our work performance and from introspection about our personal aptitudes, lifestyle preferences, needs, and values, and then use this information to formulate ideas about our work and life goals. Thus, career planning and career practice are inseparable, and we construct our careers based on such diverse influences as our parents and friends, our values and experiences as we age, and even personal emotions. What this means is that there is no one-size-fits-all approach to career planning, as evidenced by the profusion of books and guides on this topic. Rather than provide a rigid flow chart or set of steps to follow, we offer a variety of suggestions and points to consider for geographers who are themselves at different moments in their own careers. Although much of the information we present also has relevance for careers in academia, our emphasis throughout is on nonacademic careers. (See Solem, Foote, and Monk 2009 for academic career planning information.) We begin with a brief look at changes in the workplace environment that have salience for career planning, and then we turn to a consideration of the nature of the work-life fit and employment prospects for geographers.

CAREERS TODAY

Today's work environment is very different from what it was just 15 or 20 years ago, in part because of ongoing globalization, greater diversification within the labor force by ethnicity, gender, and age, growth in service occupations, and rapidly changing technology. The explosive growth of the Internet as well as other information and communication technologies signals the enduring importance of a knowledge-based economy in which creativity, innovation, and intellectual capabilities rank among the most valuable of productive assets for a firm or an organization. As a result of globalization, firms and organizations, large and small, public and private, increasingly transact business across local, regional, and international boundaries, creating a highly varied employment landscape. Alternative workplace practices such as telecommuting, flexible work scheduling, and part-time, contractual, or project-specific hiring also shape workplace culture and expectations.

These changes in the nature of work have a bearing on career planning. One way of thinking about this landscape of employment draws on the idea of a spectrum of career pathways that ranges from the "traditional career" to the "nontraditional career." Whereas job longevity with a single firm constitutes one of the defining characteristics of the traditional career, mobility among firms, agencies, or organizations is the hallmark of the nontraditional career. Neither pathway is inherently better or more lucrative than the other. Whether you seek a traditional or nontraditional career trajectory—or one that has characteristics of both—it is essential that you, as a prospective employee, not only possess important competencies, but also demonstrate a commitment to lifelong learning. Therefore, pursuit of any career demands a considerable degree of flexibility, adaptability, and resilience. Effective career planning begins when we know ourselves, when we understand the conditions of the job market, especially with respect to geography, and when we work to identify ways to mesh these factors.

THE WORK-LIFE FIT: KNOWING YOURSELF

What is your passion? Do you continually seek new experiences, or do you prefer a predictable routine? Would you describe yourself as enterprising? Imaginative? Self-confident? Are you a team player, or do you prefer to work independently? Do you function at your best in a fast-paced environment with high-pressure deadlines? Are you comfortable delegating tasks and managing personnel and work assignments on large projects? Do you enjoy public speaking? Do you have an aptitude for quantitative analysis? These are just a few questions to help you begin to think about your personal characteristics and how they might relate to the kinds of work and work environments conducive to job satisfaction. In addition, you can take a variety of self-assessments to learn more about your interests and occupational preferences, including the Myers-Briggs Type Indicator and the Strong Interest Inventory. You can arrange to take these and other assessments of personal aptitudes with a career counselor or through a campus career placement office.

Use of such self-assessments can provide helpful and often very interesting insights, but it is also important to recognize some of their limitations. For a long time conventional wisdom among career counselors was anchored to the idea that each occupation required a specific mix of traits and aptitudes. Thus, by using standardized assessment tools, people could be matched with suitable occupations. This practice not only treats people and their careers as static, in contrast to the way we have characterized them in this chapter, but it also assigns people to certain career paths and takes a narrow view of human potential. A more common perspective today incorporates the view that people adjust and adapt to the demands of their work, providing them both valuable experience and opportunities for personal growth. As an example, one of the authors of this chapter is strongly introverted and is, quite happily, a college professor with a substantial teaching load. Teaching, however, is commonly associated with extroverts, and based on some self-assessments, it would not be identified as a suitable or fulfilling occupation for such a strongly introverted person. In short, it is not always the case that our personal characteristics make us suitable or unsuitable for certain kinds of occupations. Our defining qualities, as we may see them, do not have to dictate our careers. Even if you know that you really enjoy helping others, you may find that the analytical aspects of a particular job provide ample stimulation and satisfaction, and that volunteering in the evenings or on weekends enables you to fulfill your desire to help others. Being open to different possibilities enhances your employability.

To this point our discussion has emphasized the importance of having an awareness of individual characteristics and preferences as part of the career planning process. Another significant dimension of knowing yourself involves recognizing that your personal and professional lives will tend to be closely intertwined. Therefore, family, personal, and private situations and relationships are also important considerations when you are making career decisions. Similarly, the desire to live in a particular area of the country or world, or the desire to travel as part of your work may also influence the choices—and sometimes the sacrifices—you make.

GEOGRAPHY AND CAREER PROSPECTS: KNOWING THE MARKET

According to projections produced by the Bureau of Labor Statistics (2009), almost all of the growth in employment in the next decade will take place in service occupations. Roughly half of the jobs that are growing most rapidly require a college or graduate-level degree (Bureau of Labor Statistics 2009). In addition, geospatial technology has been identified as an emerging industry that is expected to have sustained high job growth for the foreseeable future (Gerwin, 2004; Employment and Training Administration 2010).

As a geographer, you probably already know that you've chosen a very diverse field. This diversity can be both an asset and a source of frustration. On the one hand, it creates many opportunities and gives you considerable flexibility, but on the other hand it can require that you invest more time and energy searching and applying for jobs. It is especially important to keep in mind that many geographers have positions that do not use the words "geography" or "geographer." For an example, please see the profile of Valerie Moye, a geographer who works as a Green Infrastructure Coordinator. Geographers work under a broad range of job titles as diverse as GIScience Analyst, Image Processing Specialist, Geoscience Educator, Transportation Planner, and Environmental Scientist, to name just a very few.

Organizations that employ holders of geography degrees fall into three basic groups: businesses, government agencies, and nonprofits (BGNs). As Natoli (1976) suggests, geographers employed in business often serve in positions that support decision making. For example, they may be hired to identify and select optimal locations for retail stores, wind farms, or

PROFILE 1.1

Valerie Moye, Green Infrastructure Coordinator, City of Chattanooga (Chattanooga, Tennessee)

In 1969, the industrial hub of Chattanooga, Tennessee, gained notoriety for being named one of the most air-polluted cities in the United States. Today, it's a model of sustainable urban planning, and in 2009, the city established its Office of Sustainability. Although Valerie Moye was a recent college graduate who had never held a permanent full-time job, she was chosen to join the office's staff as Chattanooga's first Green Infrastructure Coordinator, thanks in large part to her proactive approach to professional development.

Valerie majored in environmental studies at Sewanee: The University of the South. While her emphases were biology and ecology, her postgraduation experiences drew her increasingly toward the discipline of geography, particularly human–environment interaction. After graduating in 2007, she served as a Peace Corps volunteer in rural Peru. Conducting outreach to promote recycling and discourage littering in a village that had no infrastructure for trash collection, she honed her patience, flexibility, and adaptability, which have served her well as she learns to navigate the intricacies of local government. She also gained tangible experience in **project management**, a "huge skill" required in her current position.

After returning home, Valerie became a GIS Fellow in the Landscape Analysis Lab at Sewanee. Working on the South Cumberland Conservation Action Plan stoked her burgeoning interest in geography and planning. To learn more about the field, Valerie scheduled an **informational interview** at a regional planning agency, and she so impressed the staff that they passed her résumé on to contacts at the City of Chattanooga. Soon she was invited to apply for the Green Infrastructure Coordinator position. "There are opportunities out there you never imagined existed," she observes.

Green infrastructure entails utilizing a community's existing network of natural resources to provide ecosystem services that are typically fulfilled by "gray infrastructure," such as pavement, pipes, and mechanical systems. Valerie notes that the benefits extend beyond obvious examples like water quality and flood control into multiple realms of urban planning, including aesthetics, pedestrian-friendly development, and street design. Interdisciplinary fields such as environmental studies and geography provide an ideal academic background for her work. She interacts with a variety of diverse stakeholders, so she relies on **strong communication skills** to produce succinct and compelling arguments to justify proposed projects. She also must demonstrate the potential impacts of projects on local neighborhoods and convey the intrinsic and economic value of ecosystems to residents. To do this, she has to display, present, and describe spatial information in a way that means something to the public, often utilizing GIS and maps.

Valerie hopes to someday pursue a master's degree, and she advocates that students take time off between their undergraduate and graduate studies. She feels **real-world experience** is essential for learning "what you're comfortable with and what you're not" in terms of your skill set and career options. As "the ultimate generalist," she's using this time to decide whether she wants to remain broadly focused or become more specialized through additional training in GIS and technology.

Valerie describes green infrastructure as an "emerging paradigm that lots of urban areas are catching hold of." Because sustainability is a "triple bottom-line proposition" that confers financial benefits, social responsibility, and ecological advantages, she observes that green infrastructure approaches are being adopted within every sector of the economy. Right now, the occupational outlook is great for geographers looking to make a living while making a difference.

—JOY ADAMS

transportation routes. Geographers employed in business also work as market researchers, statisticians, financial analysts, and software developers. Also, many private companies contract with or provide consulting services for the federal government and have positions where geographical knowledge and expertise is needed. Several other chapters in this book address the subject of working in the private sector, including Chapter 13 (consulting careers), Chapter 10 (starting a small business), and Chapter 8 (corporations and large businesses).

In the coming years, many of the U.S. federal government's 1.8 million employees will reach retirement age. Projections indicate that by 2016, nearly 61 percent of the federal government's full-time permanent employees will be eligible to retire (OPM 2008). Similar circumstances are expected in state and local governments because of the aging workforce. People with geography degrees are needed in a wide variety of federal government jobs ranging from conservationists and census specialists to managers and executives. A significant number of geographers obtain employment in GIScience, remote sensing, and cartography positions, but geomorphologists trained in watershed analysis, climatologists skilled in regional and global climate change, medical geographers proficient in public health management, and biogeographers knowledgeable about ecosystem processes and deforestation—among many others—are also needed. Geographers working for city, county, or state governments also develop careers as waste and pollution managers, urban and regional planners (such as Valerie Moye), cartographers, GIS coordinators, and educators at all levels—primary, secondary, college, and university.

Heightened concerns about homeland security and geographies of terrorism have created additional areas calling for geographical and geospatial expertise at agencies such as the Federal Bureau of Investigation (FBI), the Central Intelligence Agency (CIA), and the State Department. Increasingly, geography degree holders are serving in the United States and abroad as civilians employed by the armed forces. Incidentally, most federal jobs no longer require civil service exams. One important exception to this guideline, however, involves the Foreign Service Officer Test (FSOT). This exam is administered by the State Department three times a year (usually in February, June, and October) in selected locations and is followed by an extensive application process that can take from six months to two years to complete. For more information about public sector careers, see Chapter 6 by Bass and Quodomine (local and state government careers) and Chapter 7 by Williams, Brown, Moriarty and Wertman (federal government).

The nonprofit sector includes a vast array of scientific, educational, social justice, and charitable nongovernmental organizations (NGOs). Geography provides excellent preparation for working in nonprofit organizations that focus on issues such as environmental management, human and economic development, sustainability, gender inequalities, or historic preservation. The integrative nature of geography prepares students for positions in scientific and educational organizations that involve interaction with a diverse range of other professionals such as engineers, geologists, historians, biologists, and policy specialists. Geography also provides preparation for living and working abroad with organizations such as the World Bank or the United Nations World Food Program. You can learn much more about careers in these areas by reading Chapter 9 by Shimada and Tasch (on nonprofits and NGOs) and Chapter 11 by Mitchell and Robertson (on working internationally).

As a whole, academic geographers have been slow to recognize how well a bachelor's, master's, or doctoral degree in the discipline can prepare students for work in BGN sectors in general, and within the nonprofit sector in particular. Fortunately, several factors have converged to help correct this. First, in recent years the landscape of nonprofits has changed markedly. In the United States alone, the number of grant-making foundations—a primary source of funds for nonprofits—increased by 70 percent, and the number of public charities grew by 60 percent between 1997 and

2007 (Wing, Roeger, and Pollak 2010). Second, academic approaches have broadened so much that action research has become an important dimension of geographic practice in graduate programs. Action research is always a collaborative effort that brings together academic researchers and locals, often members of oppressed groups, to solve problems of human development. Whether conducted domestically or internationally, action research also often involves nonprofit organizations or NGOs. Third, growth in GIS and geospatial technologies has helped open avenues of communication across disciplines and in both academic and nonacademic domains. Demand exists for geographers who are well versed in the applications of these technologies as well as a mixture of qualitative and quantitative techniques to work in the nonprofit sector. For geography Ph.D. holders alone, "a national sample found that 6 to 10 years after degree completion more than 90% had successfully secured rewarding, stable work *in diverse employment sectors*" (Babbitt et al. 2008, emphasis added).

Geographers seeking to launch their careers in BGN organizations are encouraged to gain practical or supervised experience with one or more firms or organizations. Internships, whether or not they are paid positions, remain among the most valuable ways to gain "on the job" experience (for an example of a geographer who benefited from several internships, see Carmen Tedesco's profile in this chapter). Because internships provide varying kinds of work-related opportunities, it is advisable to carefully evaluate them in terms of how your time as an intern will be spent. Will you have a specific role on a project, or will you be assigned general office tasks? The faculty sponsor or internship coordinator should be able to detail the work responsibilities and expectations of the position, but contacting former interns and inquiring about the kinds of work they performed and the value of their experience with that company or agency is also recommended. If you are fortunate enough to be selected for an internship, approach your work assignment with the professionalism you would exhibit in a full-time, permanent position. See Chapter 4 by Blanchard, Carter, Kent, and Badurek for excellent advice about choosing and evaluating the right internship for you.

A good way to career plan for work with a nonprofit or NGO is simply to volunteer your time and skills to the organization. NGOs, both domestic and abroad, often face budget constraints that limit the full-time personnel they can hire or fund on projects. Offering to perform some of the less glorious tasks of routine office work or volunteering to train other staff members in the use of Global Positioning System (GPS) receivers or how to sample water quality, for example, can go a long way toward establishing a productive relationship with the organization. If you are a college student, you might consider joining the Young Nonprofit Professionals Organization for additional learning and networking opportunities.

Affiliating with an NGO can provide a valuable springboard for a career with such organizations. To affiliate means to develop a business or research relationship. For graduate students, in particular, it can be very beneficial to work with nonprofits as you do your research and fieldwork for your thesis or dissertation. This is especially important if you plan to conduct international research in a developing country. Although requirements vary from country to country, you will often need a research visa to carry out your project. In order to obtain one, you may be required to have a formal professional affiliation with a local agency. Depending on your research, you might choose to affiliate with a local NGO. If the NGO is willing to affiliate with you, a representative of the organization will write a letter that vouches for you and your planned project. Obtaining the affiliation may take a considerable amount of time, sometimes as much as a year, simply because you will have to know the personnel at the NGO and demonstrate the usefulness and feasibility of your research project to them. Acquiring this kind of affiliation can be critical to the success of your fieldwork and research. It is crucial that you take this process seriously and simultaneously recognize that having a formal affiliation brings with it important social and professional obligations. Thus, you should think about ways in which you can support the mission

PROFILE 1.2

Carmen Tedesco, Senior Spatial Planning and Development Specialist, Development Alternatives, Inc. (Bethesda, Maryland)

Carmen Tedesco became interested in biology during high school and decided to study infectious diseases in college. But she eventually realized that her interests extended beyond what she could view through a microscope: "Geography allowed a larger look at the issues related to disease. I had a combination of interests, so I wanted a broader field." Carmen now works for Development Alternatives, Inc. (DAI), a Bethesda, Maryland-based development firm dedicated to helping developing nations become more prosperous, more just, cleaner, safer, healthier, more stable, more efficient, and better governed.

Describing Carmen's job isn't easy. "My work is really eclectic," she says, "But that's one of the main benefits of being a geographer." As Senior Spatial Planning and Development Specialist in DAI's Environment and Energy Division, she collaborates with the Director of Information and Communications Technology to help integrate GIS and geospatial planning into the company's development strategies. Because she brings a combination of GIS skills and international development experience, Carmen was hired to help bridge the gap between the company's technology and programmatic sides. Much of her work involves **developing geospatial solutions** to the effects of climate change in developing countries. She also manages a U.S. Agency for International Development-funded online peer network, which she describes as a "community of practice" for natural resource management practitioners around the world.

Carmen believes that her education has been key in advancing her career. She holds a B.A. from Middlebury College and an M.A. from the University of Illinois, Urbana-Champaign, both in geography. "When you're doing any kind of work that involves technical skills, you've got to have an advanced degree," she explains. It also helps to have extracurricular experience. An undergraduate program in Bangladesh provided her with the **international perspective** necessary in her line of work. She also had several internships, which she claims are "crucial" not only for students and recent grads, but also for older job seekers who want to change careers: "You've got to know what you're talking about when you get to the job interview. Internships are a great way to get your foot in the door."

Prior to joining the DAI staff, Carmen worked for the nonprofit Academy for Educational Development (AED), where she was a member of the Climate Change Team. She wasn't hired to work as a "geographer"—her qualitative research skills and international experience got her the position. However, her job morphed toward having a more geographic focus, especially as the value of her GIS experience became apparent. Carmen believes that there is likely to be a "huge need" for employees with GIS skills in the future, as the development field is still in the early stages of utilizing the technology's full potential. In addition, she notes that social media skills, the ability to speak more than one language, and especially **critical thinking abilities** are essential for anyone looking to get a job in development. "Even for interns, critical thinking is crucial," she explains. While she has worked for both nonprofit and private development organizations, Carmen believes that the cultural differences between the two are relatively minor. "DAI is an employee-owned, mission driven company that functions very similar to a nonprofit," she explains. As for the long-term employment outlook, Carmen believes that as long as there are impoverished people in the world, there will be opportunities for geographers in the field of development.

—MARK REVELL

of the NGO, perhaps by sharing your labor, expertise, or research results. Being willing to work in tandem like this can help you build a solid reputation within the NGO community; it can also facilitate your plan to work with a nonprofit organization after you've completed your graduate degree.

We have provided some general guidelines for working with NGOs in international settings, and even though these might seem straightforward, bear in mind that working with NGOs takes time, patience, dedication, and often a considerable degree of selflessness. Sometimes there are also complex politics to negotiate, not just between the NGO and the government (from which the NGO is likely to receive some of its funding), but also between the NGO and the local community. Simultaneously, you may have to navigate linguistic and cultural differences, different expectations about gender roles, sensitive questions about research ethics, or other difficulties. Among the rewards of meeting such challenges are seeing the results of your research, gaining a wide network of professional colleagues and friends, and obtaining considerable experience and wisdom that can help you achieve your career goals.

To enhance your employability and increase your value to an organization, it can help tremendously if you are willing to take on different roles. Since many of the funds for public charities come from donations, consider making time to gain fund-raising experience. Similarly, demonstrate your willingness to collaborate by acknowledging the assistance of others. If you are involved in preparing project reports or publishing the results of a study, this can mean that you share authorship. Moreover, once you land a position, keep in mind that advancement will likely involve your taking on additional leadership roles and gaining broader awareness about other business and legal issues that affect the employer or organization. On another note, it is generally easier to move from academia into the business, government, or nonprofit sector than vice versa, primarily because of academia's emphasis on research grants, publications, and teaching. Thus, if you are considering a job in academia after working with a BGN organization, keeping active in these areas can facilitate that kind of career move. For additional information on transitioning into and across BGN careers, see Chapter 3 by Adams.

It is never too soon to start learning about different jobs or the businesses, government agencies, and nonprofit organizations that hire geographers. This is especially the case with nonprofits, because many of them are not in the "geography pipeline." For a variety of reasons, most of them don't send job notices directly to geography programs; therefore, you will need to devote more time researching them and their job vacancies. In any event, once you are hired you will likely find that you frequently seek out information about other jobs and agencies as you begin to interact with a larger number of colleagues. Canvassing the market in this way is good practice and helps you keep apprised of employment trends. It can also help you evaluate your particular job situation with respect to how employee benefits, overtime, or other workplace policies vary from firm to firm. Sometimes such information can be used as leverage for raises or other perks.

There is no single, comprehensive source of information about organizations that hire geographers, but see Table 1.1 for a list of some of the more useful job search sites. In addition, students are strongly encouraged to consult academic advisers or faculty to learn where recent graduates of their program are presently employed. It is no secret that alumni networks sometimes help a job seeker get a foot in the door. Take the opportunity to contact alumni or individuals who have the kinds of jobs you seek and arrange an informational interview with them in which you ask them questions about the educational or work experience required for the position, the nature of career advancement in their area, and social or economic circumstances that are changing how they do their job. In the span of a coffee break, an informational interview can provide very valuable job-specific information. As with any interview, you should come prepared, dress professionally, arrive on time, and follow up with a note of thanks.

TABLE 1.1 Selected Job Search Sites for Geographers

General Geography Job Sites:
http://www.geographyjobs.com/
http://www.aag.org/cs/annualmeeting/jobs_center
http://www.socialsciencesjobs.com/geography-geographer-jobs.htm

GIS Job Sites:
http://www.gjc.org/
http://www.geojobs.org/
http://careers.geocomm.com/
http://www.geosearch.com/
http://www.gisjobs.com/
http://www.giscrossing.com/
http://www.job-search-engine.com/keyword/gis-geography/

Environmental Job Sites:
http://www.ecoemploy.com/
http://www.earthworks-jobs.com/
http://www.environmentalcareer.info/

UNDERSTANDING YOUR SKILLS RELATIVE TO JOB DESCRIPTIONS

In addition to canvassing the market, another valuable strategy involves reading job announcements. In this way, you can become familiar with coursework, skills, work experience, and other qualifications necessary for specific jobs. The job announcement is an employer's basic tool for disseminating information about an opening along with an explicit list of the organization's needs (Wikle 2010). Most job announcements contain three parts: (1) basic information about the organization such as its size and mission, (2) a brief description of duties and responsibilities associated with the position, and (3) desired qualifications and minimum requirements.

Knowing what employers look for when filling specific types of positions is important for planning college coursework and identifying credentials that could enhance a professional portfolio. Another method useful in career planning is to complete a gap analysis. This is a systematic process for comparing a job seeker's qualifications with those specified in a position advertisement. Missing elements or "gaps" can be identified by making a list of skills or competencies desired or required by employers in one column and creating a second column to show the job seeker's corresponding skills and competencies. The final step in a gap analysis is comparing employer needs to personal skills and competencies. Deficiencies or gaps should be shown in a third column, together with a plan describing what will be done to address the deficiency. For example, a gap involving experience with a specific remote sensing software package could be addressed by taking an additional course or a training seminar.

When compiling lists of qualifications and competencies, it's important to look beyond duties associated with previous employment or volunteer work and focus also on more transferable skills. Such skills, acquired through a combination of our education, training, and abilities, include writing reports, managing server files, conducting interviews, and using various software packages, whether for performing remote sensing or GIS analysis, or designing maps and Web pages. Interestingly, when identifying personal skills and competencies, people are inclined to

TABLE 1.2 Top Five Skill Areas Most Frequently Cited as Needed for the Work of Employer Organizations (by Sector).

Higher Education	Government	For-Profit Company	Nonprofit Company
		Geography Skills	
• Human-environment interaction • GIS • Global perspective • Cartography • Spatial thinking	• GIS/Cartography • Spatial thinking • Spatial statistics • Field methods	• GIS • Cartography • Spatial thinking • Spatial statistics • Economic geography	• Interdisciplinary perspective • GIS • Cartography • Spatial thinking • Diversity perspective
		General Skills	
• Critical thinking • Computer technology • Creative thinking • Quantitative skills • Problem solving	• Writing • Visual presentation • Ethical practice • Computer technology • Teamwork	• Adaptability • Self-awareness • Ethical practice • Project management • Teamwork	• Many tied responses (totals too low to separate for analysis)

Note: *GIS = geographic information systems.*

Source: Solem, Cheung, and Schlemper 2008, 369. Reprinted with permission.

identify technical skills such as an ability to work with statistical data or GIS software, rather than general skills or competencies such as writing ability or the capacity to work in a team environment, even though job interviewers and hiring committees examine both sets of skills. Still, as a geographer, it is essential that you also recognize the skills that you have acquired and honed as a result of your background and training in geography. This includes the ability to think spatially, comprehend the complexities of nature–society relationships, and think holistically. For a very useful categorization of geography and general skills see Table 1.2.

PUTTING YOURSELF "OUT THERE": MANAGING YOUR VISIBILITY

A key facet of career planning today centers on making ourselves visible and capably managing that visibility. This visibility has several different dimensions, some of which are more important during different phases of the job search process. Perhaps the single most important way to make yourself visible is to create a personal Web page where you might post a résumé as well as other materials that convey your interests, qualifications, and accomplishments. Some people who prefer to work with videos create their own YouTube channels. Arguably, such websites represent the latest trend toward online portfolios. They provide a comparatively inexpensive, easily accessible, and highly visible way to showcase interactive maps that you've created or other examples of your professional creativity and competencies.

It is still important, however, that you prepare standard cover letters and résumés. Table 1.3 provides a list of some recommended guides for developing these materials. When applying for very different kinds of jobs, it is a good idea to develop cover letter and résumé templates for each type of position. These materials should be further modified to meet unique requirements outlined within an individual job advertisement. A key exception to this involves the résumés required for

TABLE 1.3 Selected Online and Print Sources for Developing Cover Letters and Résumés

Online:

Résumé tips for geography majors by Randy J. Bertolas (Wayne State College)
http://academic.wsc.edu/faculty/raberto1/geo_careers/Marketing_onesself.htm

Résumés for GIS internships by Thomas Chapman (Georgia Southern University)
http://cost.georgiasouthern.edu/geo/image4/Internship_packet.pdf

Résumé writing (Boston College)
http://www.bc.edu/offices/careers/skills/resumes.html

Résumés and cover letters (University of Kansas)
http://kucareerhawk.com/s/762/images/editor_documents/resume%20book/resume%20book.pdf

Tips on eRésumés:
http://www.eresumes.com/

Applying for a Federal Job
http://www.gpo.gov/pdfs/careers/apply/of0510.pdf

Print:

Criscito, P. 2000. *Designing the perfect résumé*. Hauppauge, NY: Barrons.

Hanna, S. I., D. Radtke, and R. Suggett. 2009. *Career by design: Communicating your way to success*. Columbus, OH: Prentice Hall.

Krannich, R. L., and C. R. Krannich. 1994. *Dynamite cover letters*. Manassas Park, VA: Impact Publications.

Whitcomb, S. B., and P. Kendall. 2008. *e-Résumés: Everything you need to know about using electronic résumés to tap into today's job market*. New York: McGraw-Hill.

federal jobs. The federal-style résumé, as it is sometimes called, requires certain specific content. The idea is that a single federal-style résumé can be used when applying to different federal jobs.

Since employers consider résumés and cover letters as writing samples—indeed, they are a kind of snapshot of the work that you do—careful editing and refining of these materials are essential. On a related note, to complement your image you should consider adopting a professional e-mail address that is based solely on your name.

Networking—the process of making contact with others for the purpose of sharing information—is another extremely valuable strategy for managing and improving your visibility. Effective networking is a skill that comes more easily to some of us than others, but it is a skill that we can all cultivate with practice. The people whom you know both personally and professionally are often a primary resource for information. The importance of networking is highlighted by estimates suggesting that more than two-thirds of job openings are not advertised publicly (Asher 2011). Information about these "hidden" positions is often transmitted among networked individuals. You can begin networking by attending regional and national geography conferences as well as more specialized conferences in your area that focus on human rights, disaster management, or green building, for example. The Association of American Geographers (AAG) and its regional divisions host a series of meetings each year; it is often possible to defer some of the costs of attending by agreeing to work at the meeting as a conference volunteer. Tina Cary offers a good deal of advice and ideas regarding professional networking in Chapter 5.

You might be wondering whether your YouTube channel or your presence on Facebook, LinkedIn, or other social media sites is enough to make you visible. The answer is probably not, at least not yet, simply because there is so much variation among agencies and job recruiters about the ways in which the Internet is used. Facebook is undeniably important, especially for the purposes of networking. Increasingly, job vacancies and company or agency information are also available via social media sites. As a word of caution, however, don't forget that employers also use social networking. There have been some cases where highly qualified candidates have not been hired because of questionable information they posted to social networking sites. If you are considering a job that requires a security clearance, as all civil service jobs with the State Department do, you might want to be more judicious in terms of whom you choose to "friend." The Questionnaire for National Security Positions, a document required as part of the security clearance process, asks about contact with foreign nationals. This is not to suggest that you cannot have friends who are citizens of other countries; it is just to point out that this is one area that is examined very closely as part of the security clearance process. Moreover, if you obtain a security clearance, you may have to modify your own social networking posts. In some jobs, for example, it simply wouldn't be appropriate to regularly detail or update your whereabouts.

Direct contact with employers is another way to become noticed. Career fairs sponsored by colleges and universities as well as the government and other organizations provide a venue for one-on-one contact with recruiters. They also provide a good opportunity to practice talking with other experienced professionals. To make a good impression, wear business attire and be prepared with several copies of your résumé. The AAG offers an opportunity for students to meet and interact with employers at the "Jobs in Geography Center" hosted at each annual meeting. In addition, green jobs fairs and meetings of local mapping professionals or other industry-specific groups provide excellent opportunities for networking. Attending engineering career fairs can also be worthwhile for geographers interested in careers related to the more technical and engineering facets of transportation planning, GIS, or environmental analysis.

The letter of recommendation constitutes another aspect of your visibility. Employers ask for letters of recommendation to verify information contained in application materials, to inquire about how an applicant gets along with co-workers, or to obtain other information useful for assessing an applicant's suitability for a position. A letter writer can be a faculty member, current or former supervisor, or other professional familiar with your academic or work experience. Even though you do not usually see the letters written on your behalf, it is still sensible to put some thought into whom you might list as a reference. A common mistake for people new to the job market is to assume that the person with the highest status credentials is the best person to ask. In academic settings, for example, it is not unusual for students to ask the department chair to write a letter for them even if they have had only a single class with that professor and in spite of the fact that they have had more meaningful interactions with other faculty members. A similar analogy from the business world is also relevant. It might not be the wisest strategy to ask the president of the company, someone you've met on just a few occasions but who has not supervised your work, to write a letter of recommendation for you. When deciding whom to use as a reference, a basic question you might ask is: "Who can best assess my qualifications for this particular job?"

There tends to be considerable variation in the protocol associated with requesting a letter of recommendation. For example, you might want to make an appointment with a potential reference to discuss the position and to explain why she or he would be a good source of information about you. If you do not know the person very well, it is acceptable to ask if she or he knows you well enough to write a helpful letter. If so, provide that person with copies of recommendation forms and basic information about you such as your résumé.

Many job applications come with standardized forms to be used for letters of recommendation, and they ask you to indicate whether you desire to waive or retain access to the letters of recommendation. If you choose to retain access to your letters, and you may have some very justifiable reasons for doing so, a prospective employer may wonder if they have received an honest assessment of you. Consequently, you might want to explain your decision, perhaps briefly in a cover letter. It is possible to do so in a professional manner without divulging highly personal information. Remember, however, that some people may simply refuse to write a letter unless you waive your access to them. Also, it is advisable to give your reference a minimum of two weeks to prepare the letter for you, although some people are likely to need more time and may expect to have three weeks or a month to prepare a letter. After the letter has been written, send a "thank you" note to acknowledge your appreciation.

CONCLUSION

Throughout this chapter we have emphasized the interconnectedness of career planning and career practice. How you plan affects what you do career-wise, and, in turn, the jobs you take shape the nature and goals of your ever-evolving career plan. We shy away from teleological views that see geographers, or anyone else for that matter, as destined for a particular occupation. However, we recognize that some people are inclined to be more active and purposive career planners than others.

It is fair to say that the value of a geography degree is now more widely recognized than ever before among nongeographers in business, government, and nonprofit organizations. Pursuing such careers may still demand considerable time, creativity, and a willingness to market yourself to others not familiar with geography. Nevertheless, because of geography's broad and integrative nature and its unique focus on geospatial methods and perspectives, geography degree holders will enter the workforce with competencies to navigate and advance within a wide range of career pathways.

ACKNOWLEDGMENTS

We would like to thank Jacqueline Vadjunec, the editors, and the three anonymous reviewers for their very helpful comments and suggestions.

REFERENCES

Asher, D. 2011. *Cracking the hidden job market: How to find opportunity in any economy*. New York, NY: Random House.

Babbit, V., E. Rudd, E. Morrison, J. Picciano, and M. Nerad. 2008. Careers of geography PhDs: Findings from social science PhDs—Five+ years out. *CIRGE Report 2008-02*. Seattle, WA: CIRGE. http://depts.washington.edu/cirgeweb/c/wp-content/uploads/2008/04/ss5-geography-report.pdf (last accessed 15 January 2011).

Bureau of Labor Statistics. 2009. *Employment projections: 2008–18*. Washington, DC: U.S. Department of Labor. http://www.bls.gov/news.release/ecopro.toc.htm (last accessed 14 January 2011).

Employment and Training Administration. 2010. *High growth industry profile—geospatial technology*. Washington, DC: U.S. Department of Labor. http://www.doleta.gov/brg/Indprof/geospatial_profile.cfm (last accessed 14 January 2011).

Gerwin, V. 2004. Mapping opportunities. *Nature* 427(22 January): 376–77.

Natoli, S. 1976. *Careers in geography*. Washington, DC: Association of American Geographers.

OED Online. 2010. "career, n." Oxford: Oxford University Press. http://www.oed.com/view/Entry/27911?rskey=pdvQH1&result=1&isAdvanced=false (last accessed 4 January 2011).

Office of Personnel Management (OPM) 2008. *An analysis of federal employee retirement data: Predicting future retirements and examining factors relevant to retiring from the federal service*. March 2008. Washington, DC: OPM. http://www.opm.gov/feddata/RetirementPaperFinal_v4.pdf (last accessed 9 January 2011).

Osipow, S. H., and L. F. Fitzgerald. 1996. *Theories of career development*. 4th ed. Boston, MA: Allyn and Bacon.

Patton, W., and M. McMahon. 2006. *Career development and systems theory: Connecting theory and practice*. 2nd ed. Rotterdam: Sense Publishers.

Solem, M., I. Cheung, and M. B. Schlemper. 2008. Skills in professional geography: An assessment of workforce needs and expectations. *The Professional Geographer* 60(3): 356–73.

Solem, M., K. Foote, and J. Monk, eds. 2009. *Aspiring academics: A resource book for graduate students and early career faculty*. Upper Saddle River, NJ: Pearson Prentice Hall.

Wikle, T. 2010. An examination of job titles used for GIScience professionals. *International Journal of Applied Geospatial Research* 1(1): 40–56.

Wing, K. T., K. L. Roeger, and T. H. Pollak. 2010. *The nonprofit sector in brief: Public charities, giving, and volunteering*. Washington, DC: Urban Institute.

2

Geography Education and Career Readiness

Joy K. Adams, Niem Tu Huynh, Joseph J. Kerski, and G. Brent Hall

IF GEOGRAPHY IS EVERYWHERE, AREN'T WE ALL GEOGRAPHERS?

The advent of Internet-based technologies has elevated the importance of geography to a level unprecedented in the history of the discipline, reinvoking the inherent tensions between the integrity of the field as a discrete academic discipline, on the one hand, and its generalist appeal on the other hand. Although this tension within geography is not new—William Morris Davis reacted to it over one hundred years ago (Schulten 2001)—geography has never been more prominent within the everyday human experience than it is today.

Geographic information and tools are becoming more and more accessible. The first decade of the 21st century has been marked by a massive growth in the volume of geographic information generated, disseminated, and consumed by governments, businesses, and members of the public. This increased spatial awareness has produced new business models centered on location-based services and enhanced global geographic knowledge.

In the current era, the ability to use geographic data, by and large, does not require formal geographic training. You and millions of others are using geography every day when operating your car's navigation system to get directions, checking in at a local business on Foursquare with your smartphone, or "reading" the landscape for clues about the individuals and groups who live in the different neighborhoods within your community. The trend toward the widespread use of geographic tools and the growth in data contributed by the public is so prevalent that it has a name: neogeography (Goodchild 2009).

So, why should you consider a career as a professional geographer? Many people have difficulty understanding what geographers do, particularly when so many basic geographic skills are part of day-to-day life. Unlike other academic disciplines (e.g., biology, engineering, mathematics), geography has a somewhat vague public image. This is not to say that the general public understands any more clearly what professional biologists, engineers, or mathematicians do for a living but only that they often see geography as a school subject rather than as a profession or career option. But, as you already realize, or will soon appreciate, geographers are employed in diverse fields and examine problems from local to global scales. Their ability to view the world within a spatial framework informed by geographic theories and concepts distinguishes geographers from other professionals in similar jobs and industries who work on related issues.

Consider the following questions as you ponder whether a career in geography is right for you (adapted from Kerski 2011):

1. Do you love maps? For thousands of years, maps have been fascinating and powerful sources of information. For example, consider how important they were during the Age of Exploration. Today's maps are not just reference sources: they are dynamic, and you can change them to suit whatever need you have or whatever problem you are trying to solve. For example, geographic information systems (GIS) combine aspects of visualization and technology.
2. Do you like to be outdoors? Geography is a field-based discipline that often depends on data collected in the field. Your fieldwork could involve observing landscape change atop a glacier, collecting sediment flows in a river, or monitoring pedestrian and traffic flows on a city street. There is no end to what needs to be mapped and analyzed.
3. Are you curious about your world? Geography allows you to investigate "what if" scenarios, to produce and test models that seek to explain and predict outcomes for many phenomena, to ask questions, and to investigate processes and their results.
4. Do you care about the well-being of your community? A career in geography can enable you to do something about local issues such as health, zoning, public services, greenways, crime, waste management, traffic, and more.
5. Do you want to make sense of data? If you think that a mountain of data exists now, just wait until next year. Geography, including geospatial technology, helps you to make sense of all of that data and to develop critical thinking skills to help you understand which data to use and which not to use.

If you answered "yes" to any of the questions above, then geography might offer you an appropriate career path. And for many reasons, geographers are particularly well positioned for success in today's job market. Gedye, Fender, and Chalkey (2004, 383) observe that:

> Geography graduates are better placed than most to enter and be effective in contemporary and future employment. Flexibility and adaptability are qualities that all geographers should develop through the study of such a varied subject discipline. There is also an emerging need in business for global knowledge and an understanding of international perspectives. This, coupled with the growing attention being paid to environmental and sustainability issues and the spatially oriented technical skills of GIS, remote sensing and geodemographics, puts geography graduates in a strong position for marketing themselves and obtaining employment.

Since the field of geography is so broad and the career options it encompasses are so diverse, it is difficult to create an accurate and representative snapshot of the employment possibilities that exist. However, it is possible to review some of the areas of employment that tend to attract geographers. In this discussion, we'll also highlight the value of geography education for preparing future professionals on different career paths.

CAREERS FOR PROFESSIONAL GEOGRAPHERS

Despite the wide application of geographic skills in many careers and industries, few of us work as "geographers." According to figures published by the U.S. Bureau of Labor Statistics (2010), only 1,170 U.S. workers bear the job title of geographer. This figure suggests that geographers are rarer than astronomers, fashion models, and animal breeders, drastically underestimating the true size and vitality of the discipline. As of 2011, the Association of American Geographers (AAG), the major professional association in the field, has almost 11,000 members participating in more than 60 specialty and affinity groups organized around major subfields. But many more people are using their geographical skills professionally, and career options are expanding at an accelerating rate. This trend reflects Hal Mooney's assertion that we are living in "the era of the geographer" (NRC 2010, ix). In 2008, for example, there were at least 145 different occupations in the Bureau of Labor Statistics database indexed to keywords such as *geography, geospatial, GIS,* and *spatial analysis* (Solem, Cheung, and Schlemper 2008). Today, many of these occupations are flagged as "bright outlook" occupations, meaning that they are expected to grow rapidly during the next several years and will potentially have large numbers of openings. This evidence supports John Frazier's observation that "geographers are employed in various positions, although few are called 'geographer'" (Frazier 1994, 29).

Many people are attracted to a career in professional geography because they want to make a difference and work for social and environmental change. Their professional experiences illustrate the range of skills that are often required to perform complex tasks such as identifying interactions and relationships between variables, managing large datasets, and selecting appropriate methods of spatial analysis (for an example of how a professional geographer uses these skills, see Steve Fearn's profile in this chapter). An education in geography not only can equip you with skills such as these, but it will also provide you with a spatial perspective on physical and human environments, as well as the opportunity to develop specialized expertise in a geographic subfield such as geomorphology, economic geography, or political ecology.

The knowledge and skills that are taught in geography curricula are transferable to and complement those used in other disciplines. They can inform any subject that relies on field data, uses statistical analysis and technology, or asks "why is this phenomenon found here?"and "how does it relate to other phenomena nearby?" Table 2.1 summarizes some fields outside of geography that benefit from the spatial perspective of geographers. These examples are part of a larger list that identifies ways of examining the world's complexity to advance knowledge of why spatial effects exist, beyond simply identifying those effects (Golledge 2002).

SKILLS USED IN PROFESSIONAL GEOGRAPHY

The Geography, Earth, and Environmental Sciences (GEES) Subject Centre at the University of Plymouth (UK) has compiled a useful list of the knowledge, skills, and competencies possessed by most geography graduates, including:

- An understanding of cultural, political, economic, and environmental issues incorporating local, regional, and international perspectives;
- Expertise in integrating, analyzing, and synthesizing information from a range of sources;
- Project management skills, such as time management, risk assessment, problem solving, and analysis gained in the course of routinely working in teams on laboratory, desk, and field research;
- The ability to generate and use a diversity of data types (text, numbers, images, and maps); and
- Flexibility and adaptability developed through working in unfamiliar and unpredictable field environments.

As this list suggests, training in geography equips graduates with a diverse combination of skills and methods as well as an integrative analytical perspective. These qualities are appealing

PROFILE 2.1

Steve Fearn, Director, Real Estate Market Research, Ross Stores (Pleasanton, California)

Like many high school seniors, Steve Fearn knew he wanted to go to college, but he had a difficult time choosing a major. After a little soul searching, he reflected on a world geography course he had taken and enjoyed as a freshman, and the wheels were set in motion. Steve made a trip to the library, checked out a book called *Geography as a Professional Field* (AAG 1966) that introduced him to location analysis and market research, and never looked back.

While he had originally envisioned a career in business, Steve found himself attracted to the **multidisciplinary nature of geography**. As a geographer, he felt that he "could study a lot of different things, and yet not be confined by being just a business major." Steve enrolled at California State University, Chico, where in his first semester the department chair asked him what he was doing there because "nobody enrolls as a freshman geography major." After taking a course in which students were asked to analyze the location of a shopping center, he found himself particularly drawn to the applied aspects of the field. Encouraged by department accolades and the advice of faculty, Steve decided to pursue his master's degree at Ohio State University.

After completing his graduate studies, Steve returned to California to begin his job search. Soon after, he landed his first job with Albertsons Grocery Stores in Boise, Idaho as a **location analyst**, a position that he describes as "grueling, but a great experience." After honing his skills at other companies such as Safeway and market research firm Thompson Associates, in 2008 Steve moved to his current position as Real Estate Market Research Director for Ross Stores, Inc., the Fortune 500 company known for its "Dress for Less" stores.

On a typical day, Steve finds himself performing many duties that he describes as "inherently geographic—looking at how things interact over distance, but at the small area or neighborhood scale." At Ross Stores, he works directly with the company's Real Estate Department to evaluate potential store locations based on a variety of characteristics, including neighborhood demographic profiles, competition, various site characteristics, and most importantly, whether or not a potential location fits within the company's corporate strategy. Steve uses **GIS** regularly in his work to calculate the impacts of proposed location sites, and he is often asked to assess how current events and global trends might affect sales projections as well as potential acquisitions.

Steve notes that being a geographer can pose challenges for job seekers, especially early in one's career, as many employers lack an understanding of the field and the skills geography majors can bring to the table: "So many people in private industry come from business, finance, and accounting backgrounds. They just don't get it." He adds that job applicants "need to learn to speak the language of people with a business background." However, he believes that **networking with other geographers** to discover which companies are already familiar with hiring geographers is key. According to Steve, the outlook is bright for geographers seeking career opportunities in his field. While rising online sales may lead to fewer "brick and mortar" stores, location analysts might become even more important as retailers face the increasingly difficult task of optimizing profit at their remaining physical storefronts.

—MARK REVELL

TABLE 2.1 Geographical Thinking Processes Applicable to Other Fields.

Geographical Thinking Processes	Relevant Fields
Understanding scale transformations	Astronomy; land use planning; engineering
Transforming representations and images	Dentistry; medicine
Understanding spatial associations	Political science
Understanding orientation and direction	Medicine; aviation; surveying; transportation/logistics management
Understanding clustering and dispersion	Demography; agriculture; health and human services management; telecommunications; epidemiology
Understanding spatial change and spatial spread	Urban planning; archaeology
Understanding densities and density decay	Retail; health services planning; city planning
Understanding spatial shapes and patterns	Landscape architecture; wildlife conservation; art
Understanding locations and places	History; military; law enforcement; planning/economic development; natural resource management; energy
Understanding the integration of geographic features represented as points, networks, and regions	Civil engineering; utility and transportation planning; hydrology; electrical and computer engineering; surveying
Understanding proximity and adjacency and their effects	Biology; environmental management; real estate
Understanding geopolitical shifts locally and globally	Political science; journalism; tourism and recreation; international studies
Using geospatial technologies	Business; transportation planning; urban planning
Understanding globalization: the interconnections between the movement of goods and people	Transportation; law enforcement; public health; international business
Awareness of biological diversity, preservation of ecosystems, and sustainability	Environmental education; conservation
Awareness of changing local and global physical environments	Agriculture; government and policy; actuarial science; climatology
Understanding social theory	Political science; economics; sociology
Recognizing diversity and differences between places	International relations; human resources; international law; international development

and relevant to today's employers. In 2007, an article published by the Association of American Colleges and Universities reported the following:

> Employers do not want "toothpick" graduates who have learned only the technical skills and who arrive in the workplace deep but narrow. These workers are sidelined early on, employers report, because they cannot break out of their mental cubicles. Broad capabilities and perspectives are now important in all fields, from the sciences to business to the humanities. (AACU 2007, 16–17)

This assessment suggests that the knowledge and skills provided by an education in geography can be attractive to many potential employers, even those whose work, at first glance, might seem to have little to do with geography. These are points Jason Welborn emphasizes in his profile.

PROFILE 2.2

Jason Welborn, Biological Science Technician, Tumacácori National Historical Park (Tumacácori, Arizona)

Jason Welborn is no desk jockey—the former backcountry crew leader, outdoor guide, and park ranger has literally taken the road less traveled. Since completing his master's degree in 2004, Jason has had a successful and exciting career that has taken him from the U.S.-Mexico border to Alaska. "I always wanted to have a job where I could work outdoors," he recalls. Today, he's doing just that as a biological science technician for the U.S. National Park Service.

Jason's first extended backpacking trip to the Appalachian Trail stimulated his early interest in geography by introducing him to new natural environments as well as different people from diverse backgrounds. Although he majored in French as an undergraduate at the University of Southern Mississippi, he minored in geography and his graduate studies at the University of Arizona focused on land management, quality of life, and political ecology, reflecting his interest in the intersection of culture and land use. After graduation, Jason secured a job as a research assistant with the Sonoran Institute, a nonprofit organization that supports conservation and restoration efforts in the western U.S. He then moved into the private sector, working as a land surveyor on the Mexican border before being laid off as a result of the recession. "I didn't want to become 'The Man,'" Jason jokes, but the uncertain economic climate encouraged him to consider federal employment. After spending the next few years in a series of short-term positions (including a summer gig as a park ranger), Jason finally found the job security he was looking for when he was offered his current post in May 2010.

A native of Mississippi, Jason now lives in southern Arizona, where he is the sole member of the natural resources staff at Tumacácori National Historical Park. Although his job title suggests that he is a biologist by training, Jason considers himself first and foremost a geographer: "As a geographer, one can do a number of different jobs without 'geographer' in the title," he points out. Despite having to learn a great deal about the biology of the park's resident species, Jason believes his background in geography prepared him well for the demands of his job, which requires a thorough **knowledge of natural systems** as well as the ability to translate that understanding to the public in a culturally sensitive manner: "The most fulfilling part of my job is being able to educate both school children and adults about the history and ecology of the park's landscape in a non-academic, tangible way." While **geographic techniques** such as mapping and GPS skills are important for his fieldwork, Jason believes **clear, succinct writing** is equally crucial: "Employees need to be able to quickly get through reports and remember them. Busy people appreciate efficiency—when you save someone time, they notice."

For anyone considering a career like his, Jason advises talking to professionals employed in the field, pursuing **internships** to get experience and build relationships, and remaining open to doing work that might seem somewhat peripheral to one's formal training. Although the availability of jobs with federal conservation agencies fluctuates with the political balance of power, he predicts that the field will continue to grow: "I'm seeing a shift in popular perception toward a recognition of the need to conserve our resources. I think there will always be a demand for people to help manage our natural systems."

—MARK REVELL

A degree program in geography provides the opportunity to develop your "geographic intelligence," building on your general knowledge and skills to become adept at choosing appropriate information, techniques, and perspectives to discover new knowledge and develop potential solutions to real world problems. Mark Fonstad (2011) has developed fifteen pillars to becoming a successful geography graduate, several of which are directly relevant to preparing for an applied career:

- *Excel in each introductory geography course.* The knowledge you gain will be useful when interacting with geographers who specialize in different areas and will also shed light on the many similarities between the subfields of geography. It can also help you describe your work to nongeographers in a clear and jargon-free manner.
- *Master geographic communication.* Geographers communicate in multiple ways including visual, verbal, spatial, and statistical formats. Learn to use these various forms of communication to address audiences of different sizes, disciplinary backgrounds, and interests.
- *Maintain a bell curve of geographic knowledge.* Know two or three topics very well so you can publish on them; know several topics well enough that you can understand published articles written on them; and know just enough about a variety of subjects to collaborate and communicate with colleagues from different disciplinary backgrounds.
- *Acquire the largest set of well-honed skills you can muster.* If you specialize in quantitative methods, make a point to learn a bit about qualitative methods (and vice versa). Update your tool set periodically to keep pace with changes in your industry and the discipline.
- *Become familiar with another discipline.* Interdisciplinary communication and knowledge are benefits in the workplace (i.e., avoid becoming a "toothpick" graduate). The ability to bridge traditional disciplinary boundaries is a hallmark of geographic thinking and an asset in today's job market.

In addition to a specialized understanding of spatial tools and concepts, geography education imparts a variety of transferable skills that are highly sought in the workplace. The term *transferable skills* refers to general skills you can apply in virtually any professional setting. Examples include time management, oral and written communication, project management, organization, problem solving, negotiation, and interpersonal skills. The value of your transferable skills should not be underestimated. Several studies suggest that professional geographers are more likely to apply general skills on the job than their specialized geographical skills (Gedye, Fender, and Chalkey 2004; Solem, Cheung, and Schlemper 2008; Hennemann and Liefner 2010). For example, Solem, Cheung, and Schlemper (2008) found that geography alumni most frequently cited general skills such as time management, communication, writing, critical thinking, and problem solving as needed "always or very often" on the job. In addition, analytical skills such as spatial thinking, interdisciplinary perspectives, GIS software use, cartography, and field methods were rated as significantly more relevant to workplace demands than theoretical or factual knowledge drawn from the respondents' specializations within their chosen subfields of geography.

Although prospective employers might not realize it (yet), geographers are uniquely prepared to meet the rapidly evolving demands of today's industries, given their big picture perspective, eye for detail, and ability to integrate and synthesize information at a variety of scales. Complementary to the development of a body of content knowledge, a well-rounded geography education typically provides training in relevant techniques and skills that open many different career pathways. For example, experience with geospatial technologies, including GIS, Global Positioning Systems (GPS), and remote sensing (RS), is an advantage in today's job market.

If you feel that you might need more training in one or more of the skills discussed above, you're not alone. Initiatives such as the AAG-sponsored, National Science Foundation-funded EDGE (Enhancing Departments and Graduate Education) Project, of which this book is one component, are increasingly documenting a perceived gap between the needs of employers and students' career preparation. Solem, Cheung, and Schlemper (2008, 370) concluded that "many geographic and general skills are in high demand, yet the curriculum offered by academic departments may not be producing those skills at a level required to satisfy that need." Research by Gedye, Fender, and Chalkey (2004) revealed that geography graduates perceived themselves to be underprepared in terms of oral presentation, leadership, and information technology literacy, some of the very skills most highly valued by employers, according to earlier studies.

This growing body of literature led AAG President Ken Foote (2010, 3) to conclude that "opportunities for geographers may be changing faster than both graduate and undergraduate programs can respond, leading to a mismatch between aspirations and education at all educational levels from BA to Ph.D." Further compounding this problem, job search skills and career preparation have been cited as areas that are underdeveloped within geography curricula. Thus, the implication is that even those students who are well prepared to enter the workforce are not necessarily well prepared to find jobs or to articulate their qualifications to potential employers (Gedye, Fender, and Chalkey 2004; Rooney et al. 2006).

This situation is not unique to the discipline of geography. A survey of U.S. employers by the Conference Board et al. (2006) determined that 93 percent of respondents considered written communication to be "very important," yet 28 percent of respondents felt that four-year college graduates entering the workforce (across majors) were "deficient" in this skill. Similarly, 82 percent considered leadership abilities to be "very important," while 24 percent rated recent graduates as "deficient" in this area. To bridge this gap, you might consider complementing your formal education with experiences that provide hands-on application of geographic concepts and skills, as we discuss in the next section of this chapter.

Many students we have worked with expect that their degree will be a ticket to guaranteed employment, and the more advanced the degree, the better the expected job and the higher the expected salary. But consider this: The students from your geography program are entering the workforce with more or less the same academic credentials as you. And you're not just competing with them for a finite number of positions; you're also competing with geography graduates from many other colleges and universities. A 2002 study by Barr and McNeilly found that recruiters typically assume that students from the same major offer more or less the same background knowledge and training. Furthermore, nearly half of the recruiters surveyed perceive that coursework is similar across majors. Given that they see degree programs as somewhat interchangeable, the recruiters indicated that they consider internships, part-time employment, and leadership positions in university organizations to be better indicators of employability than classroom experiences (Barr and McNeilly 2002). In other words, "employers are looking for much more than subject knowledge and a good degree" (GEES 2011, 5).

LEARNING BY DOING: DEVELOPING PROFESSIONAL SKILLS OUTSIDE THE CLASSROOM

Work-based learning can take many forms, including service-learning and project-based courses; placements, apprenticeships, and internships; job shadowing; and field studies (which might incorporate some kind of volunteer or work opportunity). In many cases, you can earn academic course credits for your participation. Another advantage to work-based learning is that it requires you to apply theories, concepts, and skills acquired in the classroom, deepening your understanding of course material and potentially improving your overall academic performance (Knouse, Tanner, and Harris 1999). Following are four examples of types of work-based learning experiences.

Field courses are typically one- to two-week intensive experiences, often taken during the summer to avoid conflicting with other courses. They usually provide an excellent opportunity to develop field survey techniques, to gain experience in the use of a range of geographic tools used in data collection and analysis, and to develop project management skills by executing an independent project from conception to completion. In addition, field courses expose you to real-world problems and issues similar to those encountered in the workplace.

Study abroad is typically an elective program that can range in duration from a few weeks to one or more semesters. In recent years, the option of a study abroad program has become a common feature of university education in the United States and other countries (Reilly and Senders 2009). These opportunities provide the basis for broadening traditional educational experiences, building your understanding of other countries and cultures (Amuzie and Winke 2009), and fostering a sense of global citizenship (Tarrant 2010).

Working within the department during the school year can provide you with an opportunity to develop important skills. Professors often have grants to hire students as teaching assistants for their classes or as research assistants for short- and long-term project tasks. Collaborating with a professor is often the start of a longer-term relationship that could lead to graduate school and advanced research. Whether paid or volunteer, an assistantship provides opportunities to experience geography at work while you're still in school.

Internships take different forms, but they generally provide a short-term work opportunity with an organization or company outside the university. The time frame may range from weeks to months, and the position may be paid or voluntary. Some geography programs require students (usually majors) to participate in at least one internship to graduate. In such cases, the department will often identify potential employers and help you secure a placement.

Internships are not only appropriate for students and recent graduates. They are also valuable for job seekers who are reentering the workforce, changing fields, or looking to transition from one sector to another. Internships are an important recruitment tool because they give employers the chance to observe potential job candidates before extending long-term offers (Korkki 2011). According to statistics compiled by the National Association of Colleges and Employers, organizations that use internships as a recruitment tool will, on average, draw 40 percent of their full-time hires from their pool of interns (NACE 2008). Likewise, they offer interns a good opportunity to assess whether a career with their internship employer or within a particular industry more generally is a good fit for their interests and abilities.

An unpaid internship position may be easier to secure than a paid one and can provide a meaningful professional development opportunity for geographers who are currently unemployed as a result of downsizing or restructuring. However, competition for internship opportunities of all kinds is increasing, so be proactive in your search! You can find additional advice and recommendations about internships in Chapter 4 by Denise Blanchard, Mark Carter, Rob Kent, and Chris Badurek.

Along with internships, part-time employment, volunteerism, and even travel can provide experience that will enhance your resume, meet the demands of certain employers, and help you determine the career path that's right for you. Whether you're just starting out or looking to make a change, any opportunity to apply your skills and become more familiar with working environments, organizational cultures, and performance expectations can help to set you apart from other applicants.

EDUCATIONAL STRATEGIES FOR ENHANCING YOUR EMPLOYABILITY

Although geography is an excellent field in which to develop your transferable skills and spatial abilities, interviews with professional geographers suggest that a wise strategy for enhancing your employability is to pursue academic training in a more specialized field closely related to your career interests as a way to complement your coursework in geography. For those seeking employment in the private sector, consider a minor in business or finance. Proficiency in a foreign language is becoming increasingly important across all employment sectors. For example, if you're hoping to work in international development, the ability to speak an international *lingua franca* such as Spanish, French, Mandarin Chinese, or Arabic is highly desirable. In addition to language skills, cultivate a respect for the local culture and an openness to learning about the people and places among which you will be working. Combining the integrative, interdisciplinary perspective of geography with expertise in a complementary discipline will provide you with the best of both worlds: a knack for understanding the "big picture" coupled with the ability to convey this information in ways that specialists can easily understand and relate to.

If you are a recent graduate or a professional in transition, you have a number of options for obtaining additional education, including community college courses (which are often scheduled to accommodate the needs of working students), distance learning programs, extended education or continuing education classes, and various workshops and seminars offered by professional organizations, community groups, and private companies. Occasionally, academic credit can be earned through participation in professional meetings. For example, the National Council for Geographic Education (www.ncge.org) offers a graduate credit for successful completion of a professional development program at its national conference. For some careers, a graduate degree is recommended to obtain specialized knowledge and advanced training. However, keep in mind that additional schooling is not always essential and that it is no guarantee of greater employability. Before enrolling in a graduate program for the purposes of employability, you might consider talking with professionals in your chosen field for advice on the optimal terminal degree for your career path.

LOCATING GEOGRAPHY-RELATED JOBS

As suggested throughout this chapter, geography is relevant to a wide variety of occupations and industries across the business, government, and nonprofit sectors. Geography education is inherently interdisciplinary. It encourages students to develop diverse skills, many of which are broadly transferable. Students are often drawn to geography precisely because they don't have a specific career path in mind and they appreciate the breadth and flexibility that its interdisciplinary nature affords (Gedye, Fender, and Chalkey 2004). With few jobs explicitly advertising openings for "geographers," having a clear idea of the sorts of tasks, organizations, and environments that are likely to match your skills and interests is a critical first step in beginning your job search.

You can start formulating a career strategy by examining the Jobs and Careers section of the AAG website (http://www.aag.org/careers). Here you will find information about various occupations, including descriptions, representative titles, general job duties, required qualifications, and compensation and employment trends. You can also review current job openings on websites such as Careerjet.com (which offers postings across sectors), GISJobs.com (for positions related to the use of geospatial technologies), USAJobs.gov (for federal employment), or Idealist.org (for opportunities with nonprofit organizations). Begin with umbrella terms such as *geography* or *geographer*. As you scroll through the lists and become more familiar with the sites, you might want to focus on more specific keywords, employers, and position titles that reflect your personal interests.

For an extensive list of occupations in which geographers currently work, check out the Grosvenor Center for Geographic Education's electronic archive of business cards from Texas State University-San Marcos geography alumni (http://www.geo.txstate.edu/resources/internship-jobs/business-cards.html). The business cards are grouped by specialization, including environmental geography, land use planning, geospatial technologies, geography education, business geography, real estate and construction, and interdisciplinary careers.

GEOGRAPHY IS EVERYWHERE...INCLUDING THE WORKPLACE!

Geography has never been a more relevant field. We live in an increasingly uncertain world, owing to frequent severe weather and seismic events, political shifts, global connectivity, a growing world population, and the shrinking of our resource base. At all scales and in all places, humans are confronted with geographic issues such as climate change, biodiversity loss, increased traffic flows, energy consumption, water quality and availability, natural hazards, urban sprawl, disease spread, political instability, and unsustainable agricultural practices.

Scientists from a diverse range of fields are now using geographic concepts, techniques, and theories to advance their work and to address the problems of the 21st century (NRC 2010). Employees with formal geography training will be increasingly needed to work in interdisciplinary settings, partly due to the widespread availability of a rapidly expanding body of geographic data.

We will see continued growth in the availability of spatial data generated by ordinary citizens as well as by government, industry, and nonprofit organizations in the near future. Understanding these data, their limitations, and how they can be analyzed to address real-world problems at a variety of geographic scales will only become more important. Geographers have a critical role to play in ensuring that spatial data are complete, reliable, and locationally accurate to support appropriate data use and better-informed outcomes.

Geotechnologies such as GIS, GPS, and remote sensing have transformed decision making in many job sectors. As they continue to become part of the core of modern information technology, the ability to use them will enhance your employability. Many jobs of the future have not yet been created. If an organization does not have anyone with geospatial technology skills on its staff, you may be able to create a niche for yourself. Explain how you will help save money, better meet organizational goals, offer unique ways to solve problems, and increase efficiency by applying your technological and analytical skills. With their fundamental grounding in the spatial perspective, geographers are well-poised to occupy a broad spectrum of positions across the physical and social sciences. A background in geography will enable you to grapple with today's complex and pervasive problems as well as those we will face in the years to come. Thus, geography is not only an important component of a well-rounded education; it is a discipline that is essential in the quest to achieve sustainability in the use of the planet's natural resources and to improve society and the human experience.

ACKNOWLEDGMENTS

The authors wish to thank Mark Revell and Sarah Siegel, the editors of *Practicing Geography* (Michael Solem, Ken Foote, and Jan Monk), and the anonymous reviewers for their helpful feedback and comments on earlier versions of this chapter.

REFERENCES

Amuzie, G. L., and P. Winke. 2009. Changes in language learning beliefs as a result of study abroad. *System: An International Journal of Educational Technology and Applied Linguistics* 37 (3): 366–79.

Association of American Colleges and Universities (AACU). 2007. *College learning for the new global century: A report from the National Leadership Council for Liberal Education and America's Promise.* http://www.aacu.org/advocacy/leap/documents/GlobalCentury_final.pdf (last accessed 29 June 2011).

Association of American Geographers. 1966. *Geography as a professional field.* Revised edition. Washington, DC: Association of American Geographers

Barr, T. F., and K. M. McNeilly. 2002. The value of students' classroom experiences from the eyes of the recruiter: Information, implications, and recommendations for marketing educators. *Journal of Marketing Education* 24 (2): 168–73.

The Conference Board, Corporate Voices for Working Families, the Partnership for 21st Century Skills, and the Society for Human Resource Management. 2006. *Are they really ready to work? Employers' perspectives on the basic knowledge and applied skills of new entrants to the 21st century U.S. workforce.* http://www.p21.org/documents/FINAL_REPORT_PDF09-29-06.pdf (last accessed 29 June 2011).

Fonstad, M. A. 2011. Fifteen pillars towards becoming a successful geography graduate student (Version 1.3). http://uweb.txstate.edu/~mf16/Fifteen_Pillars_1.3.pdf (last accessed 29 June 2011).

Foote, K. 2010. Supporting early career geographers beyond the academy. *AAG Newsletter* 45 (10): 3.

Frazier, J. W. 1994. Geography in the workplace: A personal assessment with a look to the future. *Journal of Geography* 93 (1): 29–35.

Gedye, S., E. Fender, and B. Chalkey. 2004. Students' undergraduate expectations and post-graduation experiences of the value of a degree. *Journal of Geography in Higher Education* 28 (3): 381–96.

GEES Subject Centre. 2011. *Geography, earth and environmental sciences employability profiles resource pack.* http://www.gees.ac.uk/projtheme/emp/empprofs.htm (last accessed 29 June 2011).

Golledge, R. 2002. The nature of geographic knowledge. *Annals of the Association of American Geographers* 92 (1): 1–14.

Goodchild, M. 2009. Neogeography and the nature of geographic expertise. *Journal of Location Based Services* 30 (2): 82–96.

Hennemann, S., and I. Liefner. 2010. Employability of German geography graduates: The mismatch between knowledge acquired and competencies required. *Journal of Geography in Higher Education* 34 (2): 215–30.

Kerski, J. 2011. Got questions? Choosing GIS as a career. Esri GIS Education Community blog. http://blogs.esri.com/Info/blogs/gisedcom/archive/2011/01/14/got-questions-choosing-gis-as-a-career.aspx (last accessed 29 June 2011).

Knouse, S. B., J. R. Tanner, and E. W. Harris. 1999. The relation of college internships, college performance, and subsequent job opportunity. *Journal of Employment Counseling* 36 (1): 35–43.

Korkki, P. 2011. The internship as inside track. *The New York Times*, 27 March 2011, BU9.

National Association of Colleges and Employers (NACE). 2008. 2008 NACE experiential education survey: Intern hiring up for fourth consecutive year. *Spotlight Online for College Employment and Recruiting Professionals* (19 March). http://www.naceweb.org/spotlight/2008/march/experiential_education_survey/ (last accessed 29 June 2011).

National Research Council (NRC). 2010. *Understanding the changing planet: Strategic directions for the geographic sciences.* Washington, DC: National Academies Press.

Reilly, D., and S. Senders. 2009. Becoming the change we want to see: Critical study abroad for a tumultuous world. *Frontiers: The Interdisciplinary Journal of Study Abroad* 18: 241–67.

Rooney, P., P. Kneale, B. Gambini, A. Keiffer, B. VanDrasek, and S. Gedye. 2006. Variations in international understandings of employability for geography. *Journal of Geography in Higher Education* 30 (1): 133–45.

Schulten, S. 2001. *The Geographical Imagination in America, 1880–1950.* Chicago, IL: University of Chicago Press.

Solem, M., I. Cheung, and M. B. Schlemper. 2008. Skills in professional geography: An assessment of workforce needs and expectations. *The Professional Geographer* 60 (3): 356–73.

Tarrant, M. A. 2010. A conceptual framework for exploring the role of studies abroad in nurturing global citizenship. *Journal of Studies in International Education* 14 (5): 433–51.

U.S. Bureau of Labor Statistics. 2010. *Occupational employment and wages, 2009.* Bulletin 2750. http://stats.bls.gov/oes/2009/may/chartbook.pdf (last accessed 29 June 2011).

3

Switching Sectors:
Transitioning into and among Business, Government, and Nonprofit Careers

Joy K. Adams

The average U.S. worker aged 39–44 has held 11 different jobs since the age of 18, and over two-thirds of these jobs have lasted no longer than five years (BLS 2010). Such high mobility within the workforce means that many of us will change employment sectors at some point in our lives. As organizations of all types are measuring their performance against the triple bottom line of people, planet, and profit, boundaries between sectors are becoming increasingly blurred, making these transitions easier and more common than ever. While government agencies and nonprofits, including many institutions of higher education, are adopting management strategies from the private sector to contend with a challenging economic climate, private sector companies and their customers are becoming more concerned about the social and environmental impacts of their business practices.

The ability to transition among sectors is a key skill for effective career development because it allows job seekers to follow or create opportunities (Everett 2007). In this chapter, I discuss opportunities and constraints for geographers within the business, government, and nonprofit (or BGN) sectors.[1] Because many geographers are employed in higher education or aspire to faculty positions, I also discuss issues to be mindful of when considering a transition from an academic position into applied work. I then present some strategies for successfully transitioning across sectors. This advice is based on the dozens of interviews with geographers I conducted to develop the professional profiles in this book, as well as my own professional experiences in different sectors and evidence drawn from related literature on career transitions.

[1]Throughout this chapter, the terms *public sector* and *government sector* are used synonymously, as are the terms *private sector* and *business sector*.

IS THE GRASS GREENER? THINGS TO CONSIDER BEFORE HOPPING THE FENCE

Geographers today are highly dispersed across sectors and industries within the U.S. workforce (Carnevale, Strohl, and Melton 2011). While every organization is unique, certain generalizations can be made about each sector that can help you decide which one is the best fit for your unique skills and interests. The following summaries explore some of the pros and cons of working in the BGN and academic sectors and discuss occupational outlooks for geographers within each of them.

Business Employment

The business, or private, sector refers to the segment of the economy composed of enterprises owned by individuals or groups. Corporations are accountable to their shareholders and operate at national or international scales. Most have clear policies and procedures, centralized operations, access to financial and training resources, and a culture that values consistency and coordination (Everett 2007). Independent businesses are privately owned by an individual or small group and usually operate at a local or regional scale. Because they are not controlled by shareholders, their values may be less profit-oriented and less risk-averse, they have more decision-making authority, and they may actively support social and community causes (Everett 2007).

Employees in private industry are likely to experience more oversight, less autonomy, and a more static set of tasks and responsibilities than their counterparts in other sectors. Although some free-spirited intellectuals may chafe at the idea of wearing a business suit and maintaining a traditional 8-to-5 schedule, a structured work environment can be a great fit for some personalities and work styles. Operations Analyst Karl Finkbeiner (profiled in Chapter 8), describes the business sector as having a "value-added, continuous improvement culture." He appreciates that his time is spent on projects that are important to the company, that he can directly measure his progress on assigned tasks, and that he can easily observe his impact on company processes and performance.

Unlike the public and nonprofit sectors, where decisions are often based on consensus and/or public input, managers in the private sector often have more authority to implement decisions and innovations (Pittinsky and Messiter 2007). Working environments are typically fast-paced, and organizational cultures value efficiency and productivity. Mid- and senior-level personnel are usually salaried, and they may be expected to put in long hours and travel frequently in order to meet their goals and deadlines (Sciencejobs.org 2011). In exchange, private sector employees are well compensated and may enjoy generous benefits and perks. However, this sector offers the least amount of job security because of the premium placed on profitability. For first-person insights on how private sector employment differs from working in the public sector, see the profile of Anne Hale Miglarese.

Despite their profit orientation, many private businesses are focusing on maximizing social and environmental benefits and minimizing negative impacts. Some companies have even developed social responsibility programs and philanthropic initiatives that address social and environmental issues. Sustainability Manager Andrew Telfer (profiled in Chapter 8) is having a far greater impact on promoting resource conservation in his position at Walmart Canada than he could have achieved in the environmental consulting career he envisioned while pursuing his undergraduate degree. The "greening" trend in the private sector is especially good news for geographers, owing to the discipline's unique position as a bridge between the social and physical sciences.

PROFILE 3.1

Anne Hale Miglarese, Principal Director, Booz Allen Hamilton (Washington, DC)

With 25 years of experience working in both the public and private sectors, Anne Hale Miglarese can be counted on to provide a solid assessment of the geospatial industry's workforce needs. She has good news for geographers: job prospects are excellent and offer a very good living for those with the right combination of skills, experience, and attitude.

Anne began her education as a marine biology major at the University of South Carolina, but she switched to geography after being introduced to remote sensing as a tool for mapping and managing coastal and wetland environments. Despite offers to work in other fields, she has never strayed from the "swim lane" of GIS and remote sensing. Although her undergraduate and graduate training both had a strong geospatial emphasis, Anne describes herself as "not much of a technologist"; her work focuses more on the "policy and people sides" of the industry. She points out that the biggest challenges facing organizations are typically not about technology, but about organizational behavior. Today, she uses this awareness to advise clients on geospatial operations, policy, and governance issues as a Principal Director with Booz Allen Hamilton, one of the oldest and largest consulting firms in the United States.

Booz Allen's clients represent the spectrum of BGN industries. Having worked in government agencies for more than 18 years, Anne specializes in the federal civil market, advising entities such as the Environmental Protection Agency and the U.S. Departments of Agriculture, Commerce, the Interior, and Defense. Her job requires a wide range of skills, including public speaking, leadership, negotiation, and project management, which she developed primarily through on-the-job experience. As someone who has worked extensively in both arenas and loved them both, Anne is quick to dispel common stereotypes about the public and private sectors, such as perceptions that government workers are less motivated and businesspeople earn much higher salaries. A major factor for her decision to transition into the private sector was her frustration with the slower pace of change associated with working within a large bureaucracy. The trade-off is that, in the private sector, "the sense of job security is only as strong as one's performance."

Anne notes that the technology needs of the public and private sectors are quite similar; however, private sector employers seek an expanded professional skill set that includes proposal writing, expertise in budgeting and estimating project costs, and project management. She suggests that prospective private sector employees take classes in information technology, industrial engineering, and organizational behavior to enhance their portfolios. She highly recommends work study and internships for all students because the public and private job markets are both becoming increasingly competitive.

Anne's ideal employee is someone who is well rounded, with outside interests, and who is comfortable with him or herself, curious, engaged, and optimistic. In addition, she looks for team players with a solid academic record, a geography degree from a good school, excellent writing skills, and multiple language proficiencies. But she especially values employees who ask "what else can I do to help?" and who want to be involved with the whole organization, not just their specific niche. "Think not only about your role or job description and doing it to the best of your ability but the mission of the organization and how you can make it better," she advises. Regardless of one's position or industry, Anne believes that the key to success is to "work hard, study hard, be very professional, and care about what you are doing."

—JOY ADAMS

Government Employment

Nearly 2 million civilians—1.8 percent of the U.S. workforce—are employed by the federal government, while state and local governments employ 19.8 million workers (BLS 2010). Geography's emphasis on addressing real-world problems and issues is excellent preparation for public sector employment, particularly at the local and state levels where much policy innovation, implementation, and bottom-line responsibility reside (Everett 2007). In addition, government employers often demand a combination of technical, conceptual, and political skills that can be hard to achieve (Everett 2007). Although this skill set can be developed through work experience, applicants new to the public sector may find the broad, integrative perspective offered by academic training in geography to be an asset.

The job title of "geographer" is more commonly used in the public sector than in other sectors, which may make finding suitable job openings and explaining the value of your geography training to some government employers less challenging. (See the profile of Sonia Arbona in this chapter and the profiles of Elizabeth Lyon and Nancy Tian posted online at www.aag.org/careers for examples.) Paid internships and part-time or seasonal positions are common and provide excellent opportunities for networking, professional development, and getting your foot in the door for permanent employment (as discussed in the profiles of Jason Welborn in Chapter 2, Dave Selkowitz in Chapter 7, and Tara Gettig in Chapter 12).

If you're seeking a "career for enhancing society and the environment" (to echo this book's subtitle), government employment will allow you to contribute to projects and decisions that will directly affect the lives of hundreds to millions of constituents. Developing and implementing policies that will affect many people and carry the weight of law involves a lot of paperwork and strict adherence to processes and procedures (Pittinsky and Messiter 2007). The rigidity of roles, slow pace of change, tendency toward consensual decision making, and competition within a pluralistic constituency can be frustrating for public sector employees, especially those who have backgrounds in the fast-paced private sector. In addition, government hiring can be slow and might necessitate completion of multiple steps and forms, particularly for positions that require a security clearance or background check.

In exchange for their persistence and patience, government employees can expect a predictable schedule of promotions and a well-defined career path, both of which afford considerable job stability (WetFeet 2003). Workplaces tend to be highly regulated, so a standard 40-hour work week is the norm and some employees are compensated for overtime. A recent compensation study sponsored by the Center for State and Local Government Excellence (Greenfield 2007) found that government workers, as a group, earn more than private sector employees. Federal employees tend to earn slightly more than their private sector counterparts, while state and local government employees typically earn slightly less. The major distinction between public and private sector wages appears to be at the senior management and executive levels, where private sector employees can potentially pull in huge salaries and bonuses.

Despite the tight job market that exists today and is expected to persist in the near future, the long-term occupational outlook for government employment is positive, given the large number of impending retirements and the creation of new roles and responsibilities. Nearly 70 percent of federal workers, 60 percent of state government workers, and 64 percent of local government workers are over 40, as compared to fewer than half of private sector employees (Greenfield 2007). Among knowledge workers (those who require specialized education, training, or skills to perform their job duties), roughly 40 percent in the federal government and more than one-third in state and local government will be eligible for retirement by 2017 (Greenfield 2007). Moreover, new initiatives, offices, and positions are being created that are particularly appropriate for geographers as governments at all levels begin to embrace the triple bottom-line paradigm. For an example, see the profile of Green Infrastructure Coordinator Valerie Moye in Chapter 1.

Nonprofit Employment

Roughly nine percent of the U.S. workforce (12 million individuals) are employed by an estimated 1.4 million nonprofit organizations, whose causes and values span the entire political spectrum (Joiner 2008). Because nonprofits explicitly strive to create a better world (as defined by their mission statements), they offer great opportunities for job seekers hoping to make a difference (Joiner 2008).

Although nonprofit salaries may be somewhat lower than those in other sectors, they weed out employees who are motivated more by earnings potential than their interest in the work itself (WetFeet 2003). Tight resources, increasing demand for services, and lean staffs may require employees to wear many hats, work long hours, and risk burnout. At the same time, lateral management structures and flexible roles provide terrific opportunities to take on more challenging responsibilities that can help you to launch or re-launch your career (Bridgespan Group 2009a).

Because more than 30 percent of the nonprofit sector's funding is soft money from government grants and contracts (Joiner 2008), the availability of career opportunities fluctuates in response to changes in the economic and political climate. Most hiring is done through networks and from organizations' intern and volunteer pools, and the process can be slow due to decentralized decision making (WetFeet 2003). Entry-level positions can be "salary-challenged," and most internships are unpaid or offer only modest stipends (WetFeet 2003). However, a looming "leadership deficit" with the impending retirement of the youngest baby boomers and high turnover among employees means that there are good opportunities for advancement (Bradley 2000; Bridgespan Group 2009a), and management and executive salaries and benefits are becoming competitive with other sectors (WetFeet 2003).

The Bridgespan Group (2009b, 2) notes that "nonprofits often look for candidates who have experience working in multi-disciplinary ways, or managing the needs of multiple stakeholders, or experience in a resource-constrained environment"; these are the transferable skills that are often emphasized in geographic research and course work. Emily Young of the San Diego Foundation (profiled in Chapter 9) hopes that more geographers will pursue nonprofit careers because their unique skill set and their content knowledge that bridges the physical and social sciences are both highly relevant to addressing contemporary social and environmental concerns. Profiled geographers working within the nonprofit sector agreed that a graduate degree (in geography or another discipline relevant to the organizational mission) is a strongly preferred qualification, as graduate studies typically provide the specialized knowledge and skills, project management and grant-writing experience, and evidence of the self-motivation necessary for success in this sector.

Nonprofit careers are particularly suitable for faculty who are transitioning into applied work. Susan Basalla and Maggie Debelius (2007, 63) observe that "the culture of nonprofits can be particularly Ph.D.-friendly, since many share values similar to those of universities." Nonprofit work can also be a good stepping stone from a faculty position to a corporate career by allowing academics to develop more organized work habits and adjust to firm deadlines (Bradley 2000). Because of their entrepreneurial spirit (i.e., learning to do a lot with limited resources), nonprofits can also provide an excellent transition for professionals moving into or out of small businesses (Bridgespan Group 2009a). Such transitions between the private and nonprofit sectors are becoming increasingly common as businesses embrace environmental and social responsibility (Everett 2007).

Academic Employment

Although doctoral programs remain predominantly focused on preparing students for faculty positions, many Ph.D.s actively seek non-academic careers, especially in geography. Nearly 25 percent of respondents to a survey of recent geography Ph.D.s work in BGN industries: 3.3 percent in nonprofits, 15.8 percent in government, and 4.6 percent in business (Babbit et al. 2008). Babbit et al. (2008) found that only 60 percent of Ph.D. geographers entered graduate school with the intention of becoming a professor, the smallest proportion observed in any of the social science disciplines they studied. For all social sciences Ph.D.s combined, 72 percent aspired to a faculty career when they began their graduate programs, but one in five of these respondents had changed career goals during the five years following graduation. Commonly cited barriers to pursuing a faculty career include push factors such as a lack of academic positions, workload demands, and the challenges of the tenure process (Golde and Dore 2001).

The decision to leave an academic position (or to abandon the search for one) is not only about push factors; there are also attractive pull factors that draw many "post-academics" into applied careers.[2] Among geography Ph.D.s surveyed five years after graduation, those working in BGN sectors earned a median annual income of $80,000 as compared to $60,000 for faculty (Babbit et al. 2008). Former faculty we profiled cited additional advantages of BGN employment, including a shorter average workweek and better work-life balance, clearer expectations for their job performance, more flexibility in deciding where to live and work, fewer constraints on partners' life choices, and greater opportunities for long-term career advancement and professional development.[3] To read about the successful transitions of some post-academic geographers into applied careers, see Sonia Arbona's profile in this chapter, along with the profiles of Emily Young (Chapter 9), Dmitry Messen (Chapter 6), and Sarah Brinegar (Chapter 7).

You do not have to completely abandon your academic interests in order to pursue a career as an applied geographer. For example, Mike Ratcliffe and Nancy Davis Lewis (profiled in Chapter 7 and Chapter 16, respectively) both serve as adjunct faculty in local geography departments. In addition, many of the post-academic profile participants observed that their current work is not much different from their previous academic employment. Dmitry Messen explained that the main difference between his former academic research and his current applied work for the Houston-Galveston Area Council is that the information he produces today is used in original planning projects rather than being published in journals. Sarah Brinegar uses her teaching skills to train her colleagues at the Department of Justice on working with GIS software and various datasets. Many of the profiled geographers continue to publish research papers and participate in professional meetings and conferences, either as part of or in addition to their work duties. Should you choose to pursue a full-time faculty position later in your career, applied work experience can be an asset. It is challenging for many professors to advise and instruct students on career development and preparation because their experience has typically been limited to academia. They often find it difficult to provide current information on job opportunities, requirements, application procedures, and employer expectations. Your insights into the world beyond the ivory tower can be invaluable to the next generation of geographers preparing to enter the workforce.

---

[2] I borrow the term *post-academic* from the excellent book *"So What Are You Going to Do with That?": Finding Careers Outside Academia* by Susan Basalla and Maggie Debelius (2007).

[3] These anecdotal accounts are supported by empirical research. In a survey by Nerad et al. (2008, 28) social science Ph.D.s working in BGN sectors reported somewhat greater satisfaction with their material resources, work-life integration, and work-family balance and just slightly lower satisfaction with the work itself than tenure-track faculty (Nerad et al. 2008, 28).

PROFILE 3.2

Sonia Arbona, Ph.D., Medical Geographer, Texas Department of State Health Services (Austin, Texas)

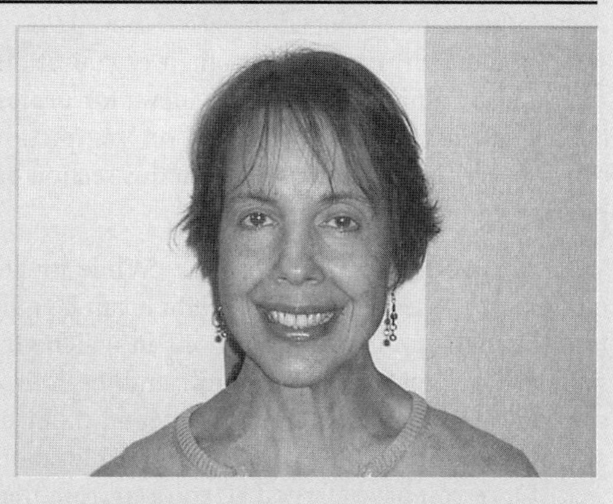

As a first-generation college student, Sonia Arbona didn't have a career path in mind when she enrolled at the University of Puerto Rico. Once she had a few courses under her belt, she decided that geography was "the closest fit" to what she wanted to do. She pursued her growing interest in public health as a graduate student at Michigan State University, researching access to health care and disease transmission in Latin America for her thesis and dissertation. "Even though we're all human, we have different risks of becoming infected because of where we are," Sonia explains. Her fascination with global patterns of disease has led to a career using GIS and geographical analysis to conduct **epidemiological studies** of HIV/AIDS and other sexually transmitted diseases in Texas.

Like many newly minted Ph.D.s, Sonia felt her graduate training had nudged her toward a faculty position. Working as an assistant professor at three universities was a positive experience overall, but Sonia wanted more time for research and family than the demands of a junior faculty position allowed. While she ultimately chose to work in the public sector over higher education, Sonia recommends that all Ph.D.s obtain some teaching experience to fully explore their employment options.

For geographers who share her passion for applied work, Sonia emphasizes the importance of **technical skills** such as cartography and GIS, stating that mapping abilities can be an "entry card" to many positions. The challenge lies in communicating that geographers have even more to offer. "Anyone can make a map," she says. "The value is not so much the tool, but the spatial perspective." Cartography is not only a way to wow her audience ("everyone likes maps!" she asserts); the maps prompt further research questions based on the geographic variations they depict. Because her findings must be clearly conveyed in a variety of formats, she also relies on **strong writing skills** and **rigorous biostatistical analysis** (the latter being a skill she developed on the job in order to communicate effectively with epidemiologists).

While she feels "very lucky" to have a job where she is able to focus on projects of personal interest, Sonia notes several trade-offs between working in a university setting and working for a state agency. As an academic, she had more freedom to develop her own research questions and directions and to focus on patterns at the global scale. As a government employee, Sonia has fewer opportunities to conduct fieldwork, her scope is limited to the state, and most of the data she uses have already been collected, so she cannot influence the methodology. On the positive side, she has access to extensive datasets that allow her to gain a broad perspective on public health issues affecting Texas, a state that includes some of the nation's largest and most diverse population clusters.

Sonia is the sole geographer in the Information and Projects Group; most of her immediate colleagues hold master's degrees in public health. However, geospatial technology is a critical tool in public health research. The Department of State Health Services has a separate GIS group that was hiring at the time of our interview, an encouraging sign for geographers in a challenging economic climate. Although not its emphasis, location is a key variable explored in epidemiology, so the integrative, spatial approach offered by geographers is very welcome in the public health arena.

—JOY ADAMS

STRATEGIES FOR SWITCHING SECTORS

Although their organizational structures, benefits, challenges, and overarching goals may differ, employers across the BGN and academic sectors are looking for a lot of the same skills and qualifications in potential hires. The profiles compiled for this book suggest that today's employers seek enthusiastic, engaged, positive candidates who can be trained in the tasks, procedures, systems, and cultures unique to their organizations over those with narrow skill sets and highly specialized training. This is great news for professional geographers, who frequently describe themselves as "jacks of all trades" or "big-picture" people, especially for those considering a career move to a different sector. In this section of the chapter, I present some key strategies for preparing to make such transitions.

1. *Investigate Your Options.* While much learning about organizational cultures and internal processes occurs on the job, it can be quite insightful to do some background research into the industry and organizations in which you seek work. Start by reviewing publications, such as the chapters in this volume and the source materials they cite. Company and industry-specific websites, newsletters, blogs, and other online resources can also yield a tremendous amount of information to inform your exploration.

Once you have learned all you can from the resources at your disposal, talking with practitioners in the industries and companies that most interest you will help answer your specific questions while also expanding your network. Professional meetings and conferences are great venues for making new contacts and increasing your awareness of various career options. For example, Esther Ofori (profiled in Chapter 8) hadn't considered a career in corporate marketing until she saw a presentation on the topic at the annual Applied Geography Conference (see http://applied.geog.kent.edu/appliedgeog/ for information). Another way to connect with geographers working in the BGN sectors is by joining the Applied Geography Specialty Group of the Association of American Geographers (see http://agsg.binghamton.edu/ for more information). Scheduling an informational interview or a day of job shadowing is perhaps the best way to learn about careers and to get one-on-one feedback on your preparation and qualifications; the process has been described as "an accelerated crash course in what it's like to be in someone's shoes" (Basalla and Debelius 2007, 84). Informational interviews are not an appropriate place to ask for a job—you must think of them strictly as fact-finding missions—but they can sometimes lead to unanticipated opportunities. For an example, read Valerie Moye's profile in Chapter 1.

2. *Identify and Communicate Your Transferable Skills.* Professional geographers have identified a common set of general skills such as time management, writing, creative and critical thinking, teamwork, public speaking, and visual communication as needed "always or very often" on the job, regardless of the specific sector in which they work (Solem, Cheung, and Schlemper 2008). As Tables 3.1 and 3.2 reveal, general skills appear to be more relevant to the work of professional geographers than specific geographic skills. This suggests that geographers with a well-rounded set of transferable skills may be able to transition between jobs and sectors with relative ease. The same is true for geographers moving into or out of academia. In their study of recent Ph.D. geographers, Babbit et al. (2008, 13) confirm that "skills in critical thinking, writing and publishing, working in an interdisciplinary context and with diverse groups of people seem to be equally important in academic and BGN sectors. Surprisingly, managing people and budgets is as likely to be important in academic as in BGN sector work." This information is encouraging for academic geographers seeking to move into applied work as well as for applied geographers who may be considering a faculty career.

TABLE 3.1 Leading General Skill Areas Used by Professional Geographers[a]

General Skill Areas	Geographers Using Skills "Always or Very Often"
Time management	91%
Writing	88%
Critical thinking	86%
Problem solving	84%
Computers and technology	83%
Creative thinking	83%
Self-awareness	82%
Visual presentation	80%
Ethical practice	78%
Information management	78%
Public speaking	75%

[a]Percentages refer to the proportion of survey respondents who reported using each skill "always or very often" on the job. The table includes only those skills cited by at least 75 percent of respondents.

Source: Solem, Cheung, and Schlemper 2008, 367.

TABLE 3.2 Leading Geography Skill Areas Used by Professional Geographers[a]

Geography Skill Areas	Geographers Using Skills "Always or Very Often"
Spatial thinking	73%
Interdisciplinary perspective	64%
GIS	58%
Cartography	53%
Field methods	52%
Human-environmental interaction	48%
Global perspective	45%
Regional geography	45%
Diversity perspective	43%

[a]Percentages refer to the proportion of survey respondents who reported using each skill "always or very often" on the job. The table includes only those skills cited by at least 40 percent of respondents.

Source: Solem, Cheung, and Schlemper 2008, 367.

Once you have identified the transferable skills most relevant to your career goals, you will want to tailor your application materials (such as your résumé, cover letter, and c.v.) to demonstrate how you have developed and applied these skills through your work experience, education, professional development, and other activities. "The key to successful career changing is learning the customs and vocabulary of the field you want to enter and then articulating your value," advise Basalla and Debelius (2007, 37). But all too often, résumés emphasize the tasks, responsibilities, and accomplishments that mattered most in one's past positions rather than what will most appeal to prospective employers. A recruiter or interviewer is less interested in what you did for someone else in the past than what you can do for him or her today, so frame your previous experience to clearly articulate how it has prepared you to meet the challenges you would encounter in the sector and organization into which you hope to transition.

Another common mistake, especially among job seekers with advanced degrees, is believing that specialized knowledge in your field is your greatest asset (Basalla and Debelius 2007). Only one-third of Ph.D. geographers in faculty positions and fewer than one-fourth working in BGN sectors reported that their dissertation topic was relevant to their jobs (Babbit et al. 2008). Furthermore, only half of BGN geographers and less than three-quarters of academic geographers with Ph.D.s in geography reported using subject knowledge from their specific subdiscipline in their work (Babbit et al. 2008). Your specializations are therefore probably less relevant to a potential employer than the transferable skills you learned through the research and writing process.

Think about your résumé and cover letter as a way to chart a course for where you're headed in your career and how you intend to get there rather than as a set of tracks showing where you've been. Be sure to demonstrate that you understand the work cultures, opportunities, constraints, and needs that characterize the employment sector and that you have the skills and abilities needed to help your potential employer achieve the organization's specific mission and goals. To do this well, you will need to perform the background research recommended above and tailor customized versions of your résumé and cover letter for each position for which you apply. You'll also want to make a compelling case for why you are the best candidate for the job among numerous applicants. One angle to explore is how you, as a geographer, can contribute a unique, integrative perspective and spatial tools and techniques that complement your transferable skills.

3. ***Be Prepared to Articulate the Value of the Geographer's Perspective.*** While geography graduates may be well prepared for success in a variety of careers, popular misconceptions exist about what geography is and what geographers do. Many BGN employers have little familiarity with the content areas or skill sets that make up our discipline beyond the ability to create and interpret maps. Therefore, it is incumbent upon each of us to clearly identify and communicate how we could contribute to an organization's goals, mission, and performance. As Sarah Brinegar advises, "[Employers] are not going to know that they're looking for you—you have to tell them."

A CEO quoted in Solem, Cheung, and Schlemper (2008, 371) explained, "We need to be able to answer the question 'Why should we hire a student with a degree in geography over a graduate from another discipline?'" Transferable skills that may be particularly well developed among geography majors, as compared to applicants trained in other disciplines, include the following:

- Expertise in integrating, analyzing, and synthesizing information from a range of sources;
- Project management skills, such as time management, risk assessment, problem solving, and analysis gained in the course of routinely working in teams on laboratory, desk, and field research;
- The ability to generate and use a diversity of data types (text, numbers, images, and maps); and
- Flexibility and adaptability developed through working in unfamiliar and unpredictable field environments (GEES 2011).

Table 3.2 presents the geographic skills most frequently cited as being used "always or very often" on the job in Solem, Cheung, and Schlemper's research (2008). Broad geographic skills and knowledge were mentioned more often than subdisciplinary or specialized tools and information. The professional geographers profiled in this book also rated overarching geographic concepts and approaches as being most relevant to their work. For example, Carmen Tedesco (profiled in Chapter 1), who works for Development Alternatives, Inc., cited her ability to focus on the intersections between cultural, ecological, environmental, economic, and biological issues as the most important skill she gained from her training in geography. Andrew Telfer compared the intersecting subfields of geography to the various facets of sustainability he encounters in his work at Walmart Canada. His ability to conceptualize these dimensions as parts of an integrated system enables him to coordinate the activities of a team of experts to ensure that their activities are contributing to the company's larger mission and objectives.

In addition, the professional geographers we profiled frequently mentioned specific technical skills that are often addressed in geography curricula as key qualifications that helped them secure their current jobs or that they seek when evaluating job applicants. Examples include geographic information systems (GIS), database skills, qualitative and quantitative analysis, and familiarity with Web 2.0 applications, including social networking. This list overlaps significantly with areas in which BGN employers stated they had "some difficulty" or were "failing" to meet their hiring needs (Solem, Cheung, and Schlemper 2008); therefore, it might be appropriate to emphasize these skills if you decide to pursue additional training or experience.

4. *Enhance Your Skill Set to Meet Employers' Needs.* Although undergraduate training in geography is an excellent platform for launching your career, more specialized academic training related to your career interests may increase your employability. Combining the integrative, interdisciplinary perspective of a broad introduction to geography with a complementary discipline or a narrower geographic subdiscipline will provide you with the best of both worlds: a knack for understanding the big picture, coupled with the ability to convey this information in ways that specialists can relate to. Further education can also increase your earning potential; a recent compensation study found that geographers holding a master's degree earned an average of 42 percent more than those with only a bachelor's degree (Carnevale, Strohl, and Melton 2011).

However, additional schooling is not always essential for certain career paths, and it is no guarantee of greater employability, especially at the doctoral level. For example, geography Ph.D. Karl Finkbeiner noted that in transportation logistics, a very applied field, academic credentials are less important than performance on the job. You'll also notice that some of our profiled geographers, such as Andrew Telfer (Chapter 8), Glenn Letham (Chapter 5), and Ken Turnbull (Chapter 5), have built highly successful careers with undergraduate backgrounds. Before enrolling in a graduate program, consider talking with professionals in your chosen field for advice on the optimal terminal degree.

If you are a recent graduate or a professional in transition, you have a number of options for obtaining additional education, including community college courses (which are often scheduled to accommodate the needs of working students), distance learning programs, extended education classes, and various workshops and seminars offered by professional organizations, community groups, and private companies. For those who have the time to invest, volunteering and internships with potential employers also provide excellent work-based learning and professional development opportunities. While some internships are only available to current students, others welcome unemployed, seasonally employed, or part-time workers. Volunteer work demonstrates commitment and interest to the work an organization performs beyond the financial rewards of paid employment. As discussed elsewhere in this volume, many employers rely heavily on their intern and volunteer pools for recruiting permanent employees, especially within the nonprofit sector where limited resources constrain recruitment activities. For more information on internships and other forms of work-based learning, see Chapters 2 and 4.

5. *Build Your Professional Network.* The importance of networking cannot be underestimated. A large majority of the geographers profiled in this book learned about their current positions through someone they knew, and many reported that they had helped someone in their network find a position. Because so few jobs advertise explicitly for "geographers," networking with other professional geographers is an excellent way to identify companies looking to hire employees with your skills and background, as Steve Fearn of Ross Stores, Inc. (profiled in Chapter 2) suggests. This information can assist you in targeting employers who understand and appreciate the value of a geographic perspective. Steve reported that about half of the staff in his division are geographers, and they continue to seek candidates with strong spatial abilities when hiring for open positions.

Expand your network to include individuals who can provide insights and information about the specific jobs, organizations, and industries you are considering. Attending professional meetings and conferences, volunteering for causes related to your interests, or joining clubs and groups are all effective ways to meet new contacts. For example, Ken Turnbull, an accredited land consultant and master networker (profiled in Chapter 5), organizes periodic networking happy hours for geospatial professionals in the Denver area. To find similar events in your local area, consider subscribing to the list serves, newsfeeds, and newsletters offered by your alumni groups and professional societies or browse sites such as Meetup.com.

Social networking sites (such as Facebook) and professional networking sites (such as LinkedIn) are also becoming increasingly important job search and recruitment tools. If you have a professional networking page or a website that highlights your academic and employment qualifications, consider providing the URL on your résumé or business card as a "landing page" where potential employers and contacts from your network can learn more about you. As Alyson Greiner and Tom Wikle point out in Chapter 1, take care to project a professional online presence. Findings from a recent survey revealed that nearly three-quarters of companies plan to increase their use of social networking sites for their initial vetting of job candidates, and over one-third had eliminated a job candidate from consideration on the basis of information found online (cited in Schuele and Madison 2010). Although online networking is increasing in importance, it is not adequate as a "stand-alone substitute" for in-person communication, explains Ken Turnbull: "Face-to-face networking, in fact, is more important now than ever because social networking is stepping away from personal contact." For more information and tips about networking, see Tina Cary's chapter (Chapter 5).

6. *Keep an Open Mind.* Many of the geographers profiled in this volume never imagined themselves in the position, career path, or organization in which they're currently working. Environmental geographer Andrew Telfer never dreamed he'd end up working for a megacorporation, Sarah Brinegar planned on a faculty career, and Armando Boniche of the *Miami Herald* (profiled in Chapter 8) thought he was leaving journalism behind when he became a geography major. However, all agree that they've ended up where they want to be, despite initially having other destinations in mind.

After speaking to scores of post-academics who now work in applied careers, Basalla and Debelius (2007, 34) observe:

> It's rare to find someone who claims to have known in advance what they would love or hate about Job A and how that experience would prompt them toward Job B and then Job C. We found that in interviewing the enormous range of people for this book, the number one answer to the question "How did you get to where you are today?" was "Serendipity."

Jim Higgins of Esri is among this group, stating that "I never, ever envisioned myself coming into sales or in managerial positions, but I grew into it." Having interviewed numerous job candidates, Jim has observed that many applicants have a very narrow view of their potential role or job title. He explained that, when offered a different position, some candidates will decline, "missing opportunities right in front of them." Like many of our profiled geographers, Jim advises that applicants think beyond the specific position for which they're applying to consider their fit within the larger organization and how they can contribute to its overall mission, not just a single functional area. (Read Jim's full profile online at www.aag.org/careers.)

Overall, the profile interviews make a strong case for being receptive to a variety of opportunities and trusting the judgment of experienced personnel to help find your best fit within an organization. In addition, be prepared to start at the bottom or in a different role than you might envision, as it could become a stepping stone to future possibilities. Even with an advanced degree and lots of work experience, entering a new sector may require you to pay your dues at first. Be prepared to set your ego aside and possibly accept a lower salary, part-time employment, more supervision, and less prestige while you learn the ropes (Basalla and Debelius 2007). The good news is that your experience and education may enable you to move up quickly once you've found your footing.

And don't be afraid to create your own niche—completing a graduate degree, designing a college course, and managing large projects will develop some of the same skills and the drive needed to become a successful entrepreneur. Anthony Morales and Kristen Carney (profiled in Chapter 10), co-founders of Cubit Planning, suggest starting small to test the waters and relying on experienced mentors to help you launch your idea. They contend that their job satisfaction is well worth their time and resources, adding that even if the business doesn't take off, you won't regret investing time into doing something you're passionate about. For more on geographers who work in small businesses, see Kelsey Brain's chapter (Chapter 10) in this book.

Despite a challenging economic climate at the time of writing, geographers are highly employable and competitive candidates for today's jobs. Not only do they bring the skills and abilities needed for all types of employment, but they also have the versatility to follow job opportunities across traditional sectoral boundaries. To reduce overhead, maintain flexibility, and maximize efficiency while attending to growing concerns for social well-being and environmental sustainability, organizations from all sectors increasingly demand employees who can address human and ecological concerns, perform diverse tasks and roles, and contribute a holistic perspective on both day-to-day operations and long-term trends and impacts. As our profile interviews and the latest research emphasize, now is an exciting time to be a geographer. I wish you luck in your job search and career!

ACKNOWLEDGMENTS

I thank Jean McKendry, Niem Huynh, Mark Revell, and Simon Engler for their comments and editorial suggestions on this chapter.

REFERENCES

Babbit, V., E. Rudd, E. Morrison, J. Picciano, and M. Nerad. 2008. *Careers of geography PhDs: Findings from Social Science PhDs—Five+ Years Out.* CIRGE Report 2008-02. Seattle: Center for Innovation and Research in Graduate Education. http://depts.washington.edu/cirgeweb/c/publications/197/ (accessed 29 April 2011).

Basalla, S., and M. Debelius. 2007. *"So what are you going to do with that?": Finding careers outside academia.* Revised edition. Chicago, IL: University of Chicago Press.

Bradley, G. 2000. Careers for Ph.D.'s in the nonprofit world. *The Chronicle of Higher Education.* http://chronicle.com/article/Careers-For-PhD-s-in-the/46376 (accessed 9 May 2011).

Bridgespan Group, The. 2009a. Bridging: An overview. http://www.bridgestar.org/Library/BridgingOverview.aspx (accessed 9 May 2011).

Bridgespan Group, The. 2009b. Frequently asked questions: Transitioning to a nonprofit. http://www.bridgestar.org/Library/FAQBridging.aspx (accessed 9 May 2011).

Carnevale, A. P., J. Strohl, and M. Melton. 2011. *What's it worth? The economic value of college majors.* Washington, DC: Georgetown University Center on Education and the Workforce. http://cew.georgetown.edu/whatsitworth (accessed 10 June 2011).

Everett, M. 2007. *Making a living while making a difference: Conscious careers for an era of interdependence.* Gabriola Island, BC: New Society Publishers.

Golde, C. M., and T. M. Dore. 2001. *At cross purposes: What the experiences of doctoral students reveal about doctoral education.* Philadelphia, PA: A report prepared for the Pew Charitable Trusts. http://www.phd-survey.org. (accessed 7 June 2011).

Greenfield, S. 2007. *Public sector employment: The current situation.* Washington, DC: Center for State and Local Government Excellence. http://icma.org/Documents/Document/Document/5046 (accessed 10 June 2011).

Joiner, S. 2008. *The Idealist guide to nonprofit careers for sector switchers.* http://www.idealist.org/info/Careers/Guides/SectorSwitcher (accessed 9 May 2011).

Nerad, M. 2007. The PhD in the US: Criticisms, facts, and remedies. *Higher Education Policy* 17: 183–199.

Nerad, M., E. Rudd, E. Morrison, and J. Picciano. 2008. *Social science PhDs five+ years out: A national survey of PhDs in six fields. Highlights report.* Seattle, WA: Center for Innovation and Research in Graduate Education. CIRGE Report 2007-01. http://depts.washington.edu/cirgeweb/c/wp-content/uploads/2008/02/ss5-highlights-report.pdf (accessed 18 October 2010).

Pittinsky, T. L., and A. Messiter. 2007. Crossing sectors: Managing the transition from private to public leadership. *Leader to Leader* 45(Summer): 47–53.

Schuele, K., and R. Madison. 2010. Navigating the 21st century job search. *Strategic Finance* 91, no. 7: 49–53.

Sciencejobs.org. Academic or industry careers? Points to ponder. http://www.sciencejobs.org/resources/pointstoponder.php (accessed 21 June 2011).

Solem, M., I. Cheung, and M. B. Schlemper. 2008. Skills in professional geography: An assessment of workforce needs and expectations. *The Professional Geographer* 60(3): 356–73.

U.S. Bureau of Labor Statistics (BLS). 2010. Number of jobs held, labor market activity, and earnings growth among the youngest baby boomers: Results from a longitudinal study. USDL-10-1243. http://www.bls.gov/news.release/pdf/nlsoy.pdf (accessed 9 May 2011).

WetFeet.com. 2003. The WetFeet insider guide to careers in non-profits and government agencies. San Francisco: WetFeet, Inc.

4

The Value of an Internship Experience for Early Career Geographers

R. Denise Blanchard, Mark L. Carter, Robert B. Kent, and Christopher A. Badurek

Broadly speaking, an internship is a work-related learning experience that introduces you as a student or recent graduate to the work environment associated with your academic education, training, and interests. An internship provides you with an opportunity to build your résumé of professional work experience, form alliances and networks in the areas of your specialization, and grasp how your coursework is preparing you to enter your chosen field.

Internships are offered in a host of career tracks, including medicine, architecture, science, engineering, law, business, and technology (The Princeton Review 2005; Liang 2006; Pollak 2007; Wise et al. 2009; Baird 2010). They may be found in private companies, nonprofit organizations (e.g., charities, environmental stewardship organizations, and research institutes), and at all levels of government. A student majoring or minoring in geography will find internships in all the subfields of geography, including geospatial technologies, environmental geography and related areas, community and regional planning, geographic education, business geography, and transportation geography, to name a few.

The goal of this chapter is to provide you with information on all the various aspects of an internship: the value of an internship for getting your career off to a good start; types and sources of internships in geography that will complement your academic studies; how to locate a suitable internship; how to maximize your internship experience; and ways for you to reflect upon and evaluate your internship experience. Internships are becoming increasingly attractive to prospective employers; for you, an internship will increase your skills, choices, and opportunities in the job market after your graduation.

THE VALUE OF AN INTERNSHIP

An internship is usually a temporary assignment of approximately three months to a year that serves as a bridge between your educational experience and future employment. Typically, internships involve work in a professional setting under the supervision and monitoring of practicing professionals. They

TABLE 4.1 The Value of an Internship Experience for Geography Majors and Minors

As an intern, you will be able to:
Obtain knowledge about the skills and abilities required for your chosen career direction.
Apply the concepts and knowledge gained in academic classes.
Acquire new professional skills.
Apply interpersonal/communications skills.
Be more motivated toward your academic studies.
Assess whether or not your academic program satisfies the expectations and goals that you have for your future career.
Acquire a network of professional contacts.
Understand the importance of roles, expectations, and behaviors in a workplace setting.
Gain confidence in your abilities.
Discern whether or not the agency or organization would be a good fit for you.
Build and/or enhance your résumé.
Enhance your graduate school application, if you choose to continue to that level.
Potentially receive an offer for an entry-level position.

may be paid or unpaid and/or taken for academic credit or noncredit. Internships offer structured educational experiences that incorporate productive work as a regular part of undergraduate and graduate curricula (COS n.d.). Specifically, an internship experience can help you:

1. determine whether your choice of an academic major and minor will translate into a meaningful career;
2. assess whether a particular industry is the best career option for you to pursue;
3. gain practical work experience related to your career goals;
4. assist you in building a record of work experience for your résumé;
5. provide you with a network of professionals who might eventually be of assistance in helping you obtain an entry-level position; and
6. offer you college credit for experiential learning.

Thus, the overall value of an internship experience is that it will provide you with opportunities to learn the roles, expectations, and behaviors of a workplace setting, and in doing so, allow you to test your strengths, interests, and problem-solving skills while acquiring valuable knowledge and contacts from associating with professionals. Table 4.1 summarizes the benefits of an internship experience for geography majors, minors, and graduate students, although the list applies to other disciplines as well.

THE IMPORTANCE OF INTERNSHIPS FOR EARLY CAREER PROFESSIONAL DEVELOPMENT IN GEOGRAPHY

You Majored in What?

In pursuing geography as your major, you can expect to face challenges that are not experienced by better-known majors such as business, medicine, and engineering. Internships prove to be exceptionally valuable in overcoming these challenges. First, geography and the role of geographers are often not well understood outside of the discipline. In her 2009 book, *You Majored in What? Mapping Your Path from Chaos to Career,* Katharine Brooks writes that when students reveal their academic majors to family and friends, ultimately THE QUESTION arises: "What are you going to do with that?" The growing number of internship opportunities available, and the increased demand for them by geography students, serve to allay apprehensions of parents and students. It sends a positive message that geography is a viable career direction with copious job opportunities.

Second, the breadth and depth of geography as a field provides myriad potential career directions. Brooks explains that the "trait and factor" logic behind THE QUESTION—which assumes a linear path between major and career—is outdated. In the early 1900s vocational researchers thought that the best way to advise for career choices was to match people's interests and skills (traits) to a range of possible vocations (factors). Today, college students often take courses and choose majors out of personal interest, not necessarily because of a specific career plan. In addition, as Curran and Greenwald (2006, xiv) write, "so much depends on interests, talent, personality—and luck." By taking advantage of an internship opportunity related to your interests within geography, or better yet, more than one internship experience, you can think nonlinearly about the various career options that geography has to offer and, thus, discern the most advantageous and compatible career direction within your interests.

Finally, the internship experience will not only assist you in adapting the "ideal" of what geographers do, but may also help employers understand the value that a geographer brings to the workplace, such as how skills and knowledge of spatial analysis can enhance understanding and decision making. Brooks (2009) points out that most employers are not familiar with the college-level study of geography, with their last memory being perhaps of a geography class in grade school that included memorizing all the states in alphabetical order. Employers may not know that a geography major takes a very interdisciplinary course of study with roots in the physical sciences as well as the social sciences, and that it equips students with skills and perspectives related to geographic information science, teaching, environmental management, urban planning, and many other areas.

Check out William Shubert's profile to see how an internship supported the professional development of one early career geographer.

THE BENEFIT OF DEVELOPING PROFESSIONAL NETWORKS

In many cases, the world of professional planners, GIS professionals, and environmental professionals is rather small. This is particularly the case within a short radius of your academic institution or within your home state. Given this reality, internship opportunities provide you with an excellent opportunity to enter the working world by helping you make key contacts in your industry in three primary ways: through (1) having a facilitator such as your direct internship supervisor, or the hiring manager of the agency; (2) attending professional development functions in your industry (Sabatino 2011b); and (3) becoming a participant in relevant alumni professional organizations (Vogt 2009).

First, internship supervisors serve as effective mentors by supplementing what you have learned in the classroom with on-the-job knowledge that can only be gained through years of work experience. In addition, internship supervisors assist you in transitioning to the workforce by serving as a key contact or professional reference when other opportunities arise. Internship supervisors who are "connectors" in their industry (i.e., those who have many professional

PROFILE 4.1

William Shubert, International Editions Coordinator, *National Geographic Magazine* (Washington, DC)

William Shubert describes his work at *National Geographic Magazine* as doing whatever he is asked to help prepare over two dozen local language editions of the iconic publication for distribution each month. A former intern, William now works as International Editions Coordinator, a job that entails a diverse and ever-changing set of responsibilities. He might be asked to write an article explaining how paleontologists determined that dinosaurs were multicolored before helping translators convey that same information in a multitude of languages, all in 200 words or less. Then it's a phone call to the Embassy of Madagascar for assistance in transliterating place-names, followed by some head-scratching to figure out how to explain "hanging ten" (the surfing term) to a Chinese audience. Factor in his emerging role as the magazine's Social Media Coordinator and a Halloween jaunt along the National Mall dressed in a Lycra zebra suit (at the behest of his editor), and one quickly gets the sense that there is nothing typical about William's typical workweek.

In a nutshell, William's job entails understanding how other cultures interpret American media. A native of San Diego, he received his bachelor's degree in 2010 from Humboldt State University in northern California. He points out that it's not the basic facts and the "cool slides" presented in his classes that have best served him in his professional endeavors, but the **global perspective** he acquired as a geography major. William credits his studies with providing fodder for the imagination, facilitating his ability to process patterns, and enabling him to **communicate effectively** in multiple modes, such as texts, maps, oral presentations, photos, and graphics—all of which directly support the mission of *National Geographic Magazine.* But he considers his international travels, including a year spent in China working a variety of jobs, as the best preparation he received for his role as a cross-cultural communicator. "Going to China forced me to go back to zero," William explains. "I thought I learned a lot in my geography and Chinese classes, but I quickly realized I knew nothing." Personally **experiencing different ways of life** showed him that that the ability to distill and convey ideas across cultural and linguistic boundaries is more important than mastering the fundamentals of a particular language or the place-name geography of a particular region. This understanding of how information can be literally lost in translation has provided William with a solid foundation for working in international media.

So, how do you get a job like William's? Like many early career geographers, he got his foot in the door as an intern. "Unlike many corporate internship programs, the purpose of our program is not necessarily to be a stepping stone to hiring at National Geographic, although it can be," explains Geography Intern Program Manager Karen Gibbs. "The objective is to further our goals in geographic literacy by providing opportunities to geography majors to **experience a work environment** in which a knowledge of geography is important. An intern, in return, brings to his or her division a geographic perspective in day-to-day operations." The program is highly competitive, so successful applicants must submit their materials on time, clearly articulate their interests, and describe their career preparation, such as other internships and job experience. (For more information, or to apply, visit www.nationalgeographic.com/jobs and look for the "Geography Intern" posting. Applications are reviewed once per year, in October.)

Although his name now appears on the magazine's masthead, William's initial application to the Geography Internship Program was rejected. Realizing that "not being good enough at the time doesn't mean you're not good enough ever," he pursued his interests and enhanced his professional portfolio through his travels, successfully reapplying from China the following year. William cites this experience as his biggest lesson learned: not withering in the face of adversity, but continuing to refine his skills in pursuit of his goals.

—JOY ADAMS

contacts or who serve as leaders in their field or professional organizations) may provide you with leads and entry to the most attractive opportunities. Thus, the choice of an internship location as well as the influence of your supervisor will play an important role in determining how effective your internship experience will be for providing access to industry networks. In some cases, students have pointed to particular supervisors who directly helped place them in compatible and enjoyable jobs (Sabatino 2011b).

Second, some internship supervisors, particularly those with a planning focus, will often bring their students to professional meetings where they may network with other professionals. Some supervisors may personally introduce you to other professionals, or it may be up to you to make use of the networking opportunity by engaging other professionals in conversation.

Third, alumni networks associated with your industry or department are useful in helping you find an internship, as well as in building your network after completion of your internship. For example, local government and GIS professional organizations frequently have meetings or networking events in which job or internship opportunities may be discussed (Sabatino 2011b).

Finally, through your internship, you will become actively engaged in professional organizations, resulting in greater opportunities for you to select internship positions and find permanent employment. In addition, an increasing number of professional and alumni organizations are using Web 2.0 applications such as LinkedIn and Facebook to build connections, strengthen networks, and manage the flow of information on employment and networking opportunities. These websites provide an opportunity for you to network virtually through e-mail exchanges and message board discussions (Cross and Parker 2004). In particular, LinkedIn provides access to online groups dedicated to very specific industries and the applied subareas of geography, including planning, sustainability, GIS, renewable energy, and environmental management (Gralla and Widman 2008; Sabatino 2011a).

TYPES OF INTERNSHIPS FOR THE MAJOR SUBFIELDS OF GEOGRAPHY

Internship opportunities that specifically ask for a geographer per se are not typical. However, if you are majoring or minoring in geography, you will discover that organizations and agencies do establish and offer internship experiences for a variety of "geography-type" positions, for example: urban planning/community development, cartography, GIS, climatology, transportation management, environmental/conservation management, professional writing/research, public information and outreach, emergency management, demography, marketing, information science, National Parks Service, and real estate. These positions may be further categorized in the major subfields of applied geography: (1) Environmental Geography; (2) Urban Planning/Community Development/Land Use Management and Regulation; (3) Geospatial Technologies; (4) Geography Education; and (5) Business Geography, Real Estate, and Location Analysis. These are the subfields where the majority of geography students seek internships. Each of these subfields will be discussed next, with examples of the jobs that interns would be observing for possible future employment (Boehm and Peters 2008).

ENVIRONMENTAL GEOGRAPHY

Environmental geography is a branch of geography that examines the spatial aspects of interactions between humans and the natural world. If you are majoring in environmental geography you might be specializing in biogeography, geomorphology, hydrology, or meteorology.

Other common areas of specialization include environmental/conservation management and assessment, hazards, and sustainable development. Your internship might involve work that calls for you to assist park rangers, environmental investigators, conservation specialists, or environmental policy specialists. Table 4.2 lists examples of internship opportunities from the website "Environmental Career Opportunities." Websites such as CampusAccess.com, ecojobs.com, and eco.org also will provide you with listings of environmental internships and careers.

TABLE 4.2 Types of Internships in Environmental Geography and Related Fields

Examples of Internship Titles
Marketing/Social Media Intern
Grassroots Internship
Environmental Education and Animal Behavior Internship
Environmental Video Transfer/Archivist
Animal Care Internship
Education Intern
Green Internship—Environmental Research
Communication and Marketing Internship
Field Internship
Public Lands Internship
Environmental Arts Internship
Education and Outreach Internship
Communications/Media Interns
Campaign Research/Development Assistant
Environmental Education and Public Outreach Intern
Sustainable Living Internships
Special Events/Marketing Intern
Resident Intern
Media Relations Intern
Educator Internship
Legislative Intern
Public Relations and Communications Internship
Communications and Social Media Intern
Summer Naturalist Interns
Journal Assistant
Environmental Education Summer Intern
Government Affairs Intern
Policy and Legislative Affairs Internship
Climate Summer Internship
Spring Outdoor Environmental Education Intern
Energy Fellow
Summer Naturalists
Wildlife Rehabilitation Internships

TABLE 4.3 Types of Internships in Community and Regional Planning, Land Use Management, and Regulation

Examples of Internship Titles
Private Land Development Intern
Real Estate Intern
Commercial Real Estate Location Analyst Intern
Municipal Government Planning Intern
Code Enforcement Intern
Neighborhood Planning Intern
County/State Government Planning Intern
Municipal Government Zoning Intern
Community Development Block Grant Intern
Public Housing Authority Planning Intern
Community Education Outreach Intern
University Facilities Planning Intern
Transportation Planning Intern
GIS Planning Intern
Surveying Intern

COMMUNITY AND REGIONAL PLANNING, LAND USE MANAGEMENT, AND REGULATION

In addition to fostering a sense of place for a geographic location, planners with backgrounds in geography are also aware of spatial relationships and patterns that both define a community and influence its various capacities, including housing, business development, public health and safety, disaster management, transportation, and recreation. Planning internship opportunities may be found at all levels of government, as well as in global corporations, architectural firms, and other businesses. Table 4.3 lists types of internships if you have a career interest in these subfields.

GEOSPATIAL TECHNOLOGIES

Geospatial technologies comprise the fastest growing area of geography, and include geographic information systems (GIS), Global Positioning Systems (GPS), and remote sensing, all powerful tools used to store, analyze, visualize, and present spatial information. Together with appropriate cartographic techniques and principles, geographers are using geospatial technologies to better understand the interactions of various factors across space, including population distributions, traffic movement, land use and availability, environmental hazards, soil types, and vegetative cover. Any phenomena that can be tied to a geographic location (georeferenced) may be analyzed spatially, which means that geospatial tools can (and often should) be used in every aspect of the practice of geography. If you are specializing in areas of geospatial techniques, you are positioned comfortably for internships and jobs well beyond the traditional bounds of the discipline.

Internships as well as career and job opportunities in geospatial technology are available in both the public and private sectors, in fields ranging from environmental conservation to resource speculation to health and safety administration, computer science, resource management, and

TABLE 4.4 Types of Internships in Geospatial Technologies

Examples of Internship Titles
Public Sector GIS Interns—various titles at the local, county, state, and federal level
Private Sector GIS Interns—various titles
Land Development Intern
Map Analyst Intern
Floodplain Mapping Intern
Satellite Imagery Analyst Intern
GPS Technician Intern
Cartography/Map Design Intern
Facilities Infrastructure Inventory/Management Intern

education. Government agencies at local, state, and national levels are among the largest providers of internships for students trained in geospatial technologies. Table 4.4 lists internship titles that may be of interest to you in these career fields; however, the list is by no means exhaustive.

GEOGRAPHY EDUCATION

Geography education extends beyond the boundary of teaching in a classroom setting. Museums, nonprofit organizations, and government/quasi-government agencies seek interns to assist in developing, implementing, and conducting public outreach programs, educational seminars, and workshops as well as in making contributions to educational policy at the local, state, and national levels. For example, as an intern, you might assist an education specialist working for a government agency charged with the general supervision, maintenance, facilitation, and development of age/grade-level-appropriate programs that comply with educational standards. Table 4.5 lists some of the internship opportunities to consider if you enjoy teaching in informal venues.

TABLE 4.5 Types of Internships in Geographic Education

Examples of Internship Titles
Museum Curator Intern
Museum Tour Guide
Visitor Assistance
Historical Preservation Intern
Event Presenter or Speaker
Environmental Education Intern
Education Conference/Workshop Intern
Writer for Environmental Education Materials
Developer of Geography Education Materials
Geography Teacher (informal setting)
Media Intern
Research Assistant Intern

TABLE 4.6 Types of Internships in Business Geography, Real Estate, and Location Analysis

Internship Titles
Location Analysis Intern
Real Estate Development Intern
Business/Economic Development Intern
Shopping Center Development/Management Intern
Aerial Photography Interpretation Intern
Mapping Intern
Demography/Population Geography Intern
Communication Network Intern
Global Trade Intern
Municipal Facilities Research Intern

BUSINESS GEOGRAPHY, REAL ESTATE DEVELOPMENT, AND LOCATION ANALYSIS

Geographers are increasingly in demand in the business community because of their abilities to synthesize spatially diverse information, to uncover spatial patterns, and to solve spatial problems using 21st-century geospatial technological tools such as GIS, GPS, and remote sensing. If you are pursuing business geography/real estate/location analysis, then you are learning the importance of location and spatial distribution of economic activities. Your focus may be on the spatial dynamics of trade, transportation, migration, capital flows, and communication networks.

By majoring or minoring in business geography, you will obtain diverse knowledge and skills related to location analysis. Combined with geospatial tools, you, in your internship, will play a key role in a vast array of decisions for the public and private sectors, such as the placement of municipal utilities, corporate headquarters, or transportation routes. You can also hone your reasoning and evaluative abilities, as well as apply quantitative, qualitative, and geospatial tools for decision making. In addition, your knowledge of cultural geography will help you land internships related to trade, policy, law, and business relations. Table 4.6 lists internship opportunities to consider if you are interested in business geography, real estate/economic development, and location analysis.

CONSIDERATIONS FOR CHOOSING AN INTERNSHIP

The "Right" Internship

Finding and choosing the "right" internship is important. If you are seeking an internship, you should begin by asking yourself: "Just what are my career goals?" Perhaps the most important question that you, as a prospective intern, might ask is: "What kind of work do I want to pursue as a career after I complete my university degree program?" While it is usually difficult to answer this question, it is a crucial exercise. At the very least, it helps to provide you with some personal guidelines about the kind of position as well as the employment sector where you should begin your internship experience—and, you should follow your instincts and interests as much as possible in this regard. Professional careers may last decades and often endure for the balance of your working life; thus, choosing the right internship will have a major impact on launching your career properly. The bottom line is this: first and foremost, choose an internship that complements your professional and career interests.

In general, you should take a number of considerations into account when deciding on which internship opportunity to pursue. For many, the number one concern is whether the internship is paid or unpaid.

Paid or Unpaid?

All things being equal, it would seem at face value that paid internships are more valuable than unpaid ones. All things are not always equal, however, and it may be that an unpaid internship will be far more useful than a paid one. For instance, a paid internship in an industry or employment category that is not one in which you see potential for a career, is essentially no better than a part-time job flipping burgers in a fast-food joint. It pays, but it brings you no closer to advancing your career. Similarly, even an internship in a business or government agency that fits your career and professional goals may not be very meaningful if your duties are mostly photocopying, data entry, answering the phone all day, and other repetitive and unchallenging tasks. In general, an unpaid internship that is: (1) professional and career-oriented, (2) exposes you to a dynamic work environment, and (3) includes challenging and intellectually demanding work related to your field of study will likely be a far better choice than the paid internship that does not provide these advantages, or does so in far lesser measure than the unpaid alternative. Because many internship experiences last only one semester, accepting an unpaid internship in your junior year of university study could later qualify you for a paid internship during the summer or the next academic year when you are a senior or in a graduate program. So it is useful to bear in mind that in some circumstances, the benefits of an unpaid internship, even in the short run, may far outweigh the benefits of the paid alternative.

In addition, the focus on being paid may be a short-sighted concern when you are weighing the benefits of potential internship opportunities. For example, paid internships in distant locations often carry large costs related to housing and transportation to the internship site. When considering the financial aspects of an internship, don't forget that if you live in a rented apartment close to your college or university, or at home with your parents, it may be better to pursue an unpaid internship. In addition, if you pursue an internship in your hometown, you may be a preferred applicant by many potential agencies that seek to provide opportunities for students from the local community. If distance is not a concern, organizations such as the Student Conservation Association (SCA 2011) provide access to many internship opportunities for students by brokering internships at sites such as the National Park Service, Bureau of Land Management, and the United States Forest Service. You might receive low pay or a small stipend for your duties, but this will be balanced by free housing or perhaps reimbursement for travel to the internship site. Internships found through SCA are popular with geography students because they sometimes focus on outdoor data collection duties in exotic locales.

WHAT KIND OF WORK IS OFFERED?

Another common concern that you might have is the nature of the work involved in an internship, such as the trade-off between indoor and outdoor duties and the amount of analytic work. Many students, particularly those with strong interest in physical geography, tend to look for internship opportunities that involve gathering field data, collecting GPS points on natural features, or engaging in physical activities related to stream restoration or trail maintenance. However, these positions may limit opportunities to engage in professional activities such as data management, spatial and environmental analysis, and contributions to final reports. You may also experience the opposite at office internship settings where the entire experience may be limited to data entry or tasks such as digitizing GIS data or georeferencing scanned photographs or historic documents. In this case, you may be engaged in a work experience in a more "professional setting," but you will also want to look for opportunities to develop the critical thinking and analytical reasoning skills that many professional organizations have argued are needed for meeting the demands of the GIS/IT and applied geography workforce (NRC 2006, UCGIS 2006). Appropriate mentoring at the internship site and some variety in duties is a key aspect for you to consider in selecting an internship to maximize your educational and professional development according to your own particular interests.

WHEN IS THE BEST TIME TO PURSUE AN INTERNSHIP?

A third major consideration is the timing of your internship placement. Generally, students may engage in internships for credit or not for credit at any time in their academic career. However, employers generally prefer to have interns with a set of skills and abilities that will enable them to perform work helpful to the organization. Since most geography undergraduate students begin taking higher level courses in their junior year, most internships that require full- or half-time work commitments would be conducted during the summers after your junior year, after your senior year in some instances, or during a graduate program. For example, some academic institutions encourage students to pursue an internship experience at the end of their senior year because they then have the highest levels of skills and knowledge to be effective at the job site. This timing also might help you more easily make the transition from your student environment to life in the workforce. In some cases, it may also help you procure a full-time job at the internship site. In many instances, employers will "try-out" summer interns to gauge their skills and abilities and potential contributions as employees before offering full-time employment. However, employment is not always a guarantee—many agencies may explicitly state that they do not hire interns after completion of the assignment.

FOR COURSE CREDIT OR NOT? ADVANTAGES OF AN INTERNSHIP FOR ACADEMIC CREDIT

Though uncommon, some internship sponsors may require that you be registered in an internship-for-credit program. You may certainly look into pursuing and performing an internship on your own, but you should also consider the following advantages to enrolling in an academic for-credit internship course:

1. An internship course counts for credit toward your degree and graduation (for example, for three credit hours).
2. Other than agreed upon meetings with your internship coordinator, there are no class meetings required for an internship. Furthermore, you and your sponsoring organization's supervisor determine working hours and schedule.
3. An internship "course" is a great line on your official college transcript; it adds credibility to your experience.
4. You will have priority assistance in finding an internship from your internship coordinator.

5. Your internship experience has accountability. You are getting graded by both your internship supervisor (performance evaluation) and internship coordinator (instructor of record).
6. To make a habit of recording and documenting what you do at work, you will probably be required to keep a daily work-log and prepare an assessment report on your own. This can help you become more self-motivated and responsible.

HOW TO LOCATE AN INTERNSHIP—SEARCH WIDELY

Because geography is a vast field, the range of possible internships is also large, yet the number of internships that might include the word *geography* or one of its derivatives in its title is decidedly limited. This is in no way a serious impediment to geography students successfully landing excellent internship opportunities. Government agencies, businesses, and nonprofits seek interns under a wide range of job titles. It will behoove you to be aware that, while the title may not invoke *geography*, geographers have many skills and intellectual perspectives that allow them to fill these positions. In seeking an internship, you should read job postings and advertisements carefully, realizing that the word *geographer* or *geography* will rarely appear—pay special attention to the description of duties and skills required. It is often here that you can identify good internship prospects that may be good career builders.

THE SEARCH PROCESS

So, how to begin? First, whether or not you are one of the lucky ones whose department provides assistance in the internship search, you need to accept the fact that a good search means you will have to "put yourself out there." You will need to find your courage, put on your thick skin, and prepare to be insistent and patient at the same time. At times you will need to be a bit aggressive about pursuing opportunities and getting your name and résumé in front of potential internship sponsors. It is also important to realize that, while looking for an internship is much like looking for a job, it is also distinctly different in many ways. A significant difference emanates from the fact that many are not formally advertised and, instead, are filled through informal networks. In any case, a search for an internship may be facilitated through careful review of job boards (e.g., www.indeed.com or www.monster.com); websites for governments, businesses, and nonprofits; and employment announcements. In addition, some large corporations, government agencies, and even professional associations may sponsor internship programs and actively seek student applicants throughout the year. In these cases, enrollment in a specific course of study or at least in an approved university-level internship course may be required.

Many internships, however, are not filled through advertisements or postings on websites. They are filled through formal and informal inquiries and contacts. This is where you cannot be too bashful. Even from the beginning of your internship search, you should pursue and cultivate informal and formal means of landing a position. Talk with your professors and fellow students, tell them what kind of an internship you are seeking, and ask if they have any suggestions. Professors often know local individuals and organizations that use interns and may be able to suggest where to start. Your fellow students may not have the same breadth of contact with professionals, but still may be knowledgeable about local internship opportunities or may simply have some useful suggestions for where to look. You may also want to talk with your parents, their friends, and your relatives about your career plans and your search for an internship. These individuals may work in the same employment sector that interests you, or know someone who does.

"Cold calling"—that is, simply picking up the telephone and calling without any kind of introduction or prior contact—or simply showing up unannounced at a government office, business, or nonprofit might also work. Some students do obtain an internship this way; however, it is not easy, and not for the faint-of-heart. In most cases, rather than a cold call or an unannounced visit, it is probably best to use an incremental approach. Once you have determined where you would like to work as an intern, try to identify the professionals in the office or department that interests you. Send them a brief e-mail identifying yourself, the university you attend, your major, and your academic and professional interests and goals. Attach a résumé.

If you are fortunate, you will be in a department or college where faculty and staff have a long tradition of sponsoring internships and who have a broad network of local contacts to draw on. In these cases, the internship program serves as a clearinghouse bringing students together with those in businesses, government agencies, and nonprofit organizations seeking interns. There may be a full- or part-time faculty member or a member of the professional staff who serves as an internship coordinator. These individuals have a wealth of knowledge and can often facilitate your internship search.

BE PROACTIVE

A proactive approach to finding an internship, as described by Richard Nelson Bolles in his seminal career-planning guide, *What Color Is Your Parachute?*, begins with determining your career goals, then researching organizations in which people have similar career goals, and finally, finding and connecting to those people with personal inquiries, informational interviewing, job shadowing, and networking strategies. Internships that you obtain proactively by establishing professional relationships with potential internship sponsors tend to be the most satisfying and successful experiences.

Starting places for you to research internship-sponsoring organizations include alumni directories, career fairs, archived job descriptions, and keyword searches on the Internet. You should also utilize campus Career Services Department staff to help identify potential internship-sponsoring organizations. Table 4.7 lists a variety of ideas for locating internship opportunities.

TABLE 4.7 Sources of Internship Opportunities
Departmental alumni. Many obtained their start as interns themselves and often will send a request to a geography department for an intern.
Public, private, and nonprofit organizations that have previously had success with hiring students as interns or somehow heard about a particular department, often will send the department a request for an intern.
Recurring "contract internships" with other state agencies.
Recurring state and federally managed internship programs.
Recurring on-campus internship opportunities.
Searching on the Internet for posted internship opportunities.
The result of a student inquiry/informational interview.
The result of a successful *job shadowing* experience.
Family member, friend, or other "network."
A chance opportunity ("the right place at the right time"). Be prepared to take advantage of serendipity.

Students are often insecure about the qualifications or skills that they might need for an internship. Although sponsors often look for specific skills and abilities, they generally want to hire interns that are thoughtful, intelligent, and hardworking, and who bring an enthusiasm and willingness to learn to the position. It is difficult to exaggerate the importance of communicating your enthusiasm and interest when seeking an internship.

DUTIES AND EXPECTATIONS: LAYING THE GROUND RULES

It is important to get a clear idea of your internship duties. Keep in mind that written internship descriptions may be vague or general, and it is sometimes the case that your internship sponsor is not altogether clear on how your duties might evolve. Nevertheless, when the opportunity arises—when interviewing for the position, or later if the internship is offered to you—you should make every effort to clarify what will be expected of you. Potential misunderstandings and problems can often be avoided through a written agreement signed by you and the sponsoring agency that spells out your responsibilities. If your internship is to be used as part of a formal *for-credit* experience, then the faculty supervisor should also be included as a participant and signatory of this agreement.

MAXIMIZING THE VALUE OF YOUR INTERNSHIP DURING AND AFTER THE PROGRAM

Upon accepting and beginning an internship, you should make a point to keep track of your work accomplishments and the amount of time spent interacting with your internship supervisor and other staff. In addition, always conduct yourself with a professional work ethic and build rapport with your internship supervisor and co-workers. Often students do not consider the effort it takes by employees to supervise interns. In many cases, supervising interns is an additional duty performed by busy professionals who may or may not have volunteered for the duty. Those employees who seek out or volunteer to supervise you often do so out of interest in assisting fellow university alumni, developing young professionals in their industry, or assisting their organization by providing outreach opportunities and/or completing work tasks. In addition to expected professionalism, you should assist your supervisor by being respectful and patient with receiving duties and feedback on performance.

Keep in mind that you are receiving the latest training in technology and states of knowledge in your field and that not all employees may be as knowledgeable about GIS and IT or be as passionate about their current position as you are. If you feel that you might be doing more advanced-level work, supervisors often appreciate a proactive approach where you have delivered products or work relevant and useful to the organization that was not directly supervised. For example, GIS and cartography interns are often capable of creating many kinds of maps or performing analyses that may be useful to organizations without having been guided directly to complete these tasks. A student who waits around to be told what to do by supervisors usually reports greater dissatisfaction with his or her internship experience than a more actively engaged and proactive intern.

Build a good relationship with your supervisor and co-workers by demonstrating a positive attitude and good work ethic. Many employers have reported that they would much prefer a pleasant, enthusiastic intern with good technical skills to an exceptionally skilled student with an unpleasant demeanor.

You may also enhance your internship satisfaction by focusing on developing your skills in teamwork, project management, leadership, and verbal communication in addition to your technical practice while on the job. Table 4.8 provides important tips for ensuring that your internship is a successful experience.

TABLE 4.8 Keys to Student Intern Success on the Job
Be professional.
Be respectful and patient.
Be proactive.
Give some value to the organization.
Be mentored.
Help your internship supervisor.
Practice soft skills as well as academic skills.
Reflect on duties and experience often.
Keep track of progress and accomplishments.
Create a professional portfolio or other evidence of work practice.
Build connections and a network.

EVALUATING AND REFLECTING ON YOUR INTERNSHIP EXPERIENCE

By regularly logging your work tasks, accomplishments, and other tangible evidence of success while on the job as well as by keeping maps, graphics, planning documents, or other products created in a professional portfolio, you will ultimately be able to evaluate and reflect upon your experience. You will also be able to provide your faculty/staff supervisor with a record of your performance. In addition, a summary letter of performance or evaluation check sheet should be solicited from the internship supervisor prior to completing your exit interview. This evaluation will provide you with valuable feedback on strong points and areas of improvement, though in practice most supervisors tend to limit negative feedback in written documents. See the profile of Mark Barnes for an example of how reflecting on an internship experience can clarify one's professional interests.

If you are taking an internship for academic credit, you may receive feedback on your internship performance from your internship coordinator through e-mail communication, site visits, commentary on your final portfolio, or in the form of a debriefing on any discussion between your internship supervisor and faculty/staff supervisor. Table 4.9 presents questions that you might

TABLE 4.9 Reflecting and Reporting on Your Internship Experiences
What were your basic duties?
Did you enjoy the work?
Did you enjoy the area/region/environment you were working in?
Did you handle any important assignments that were not part of your basic duties?
Do you have a better understanding of what kind of work you'd like to do after graduation?
Did the internship meet your expectations?
Did the internship help you develop work skills in written/oral communication, responsibility, leadership, work ethic, time management, teamwork, and taking direction?
Do you think the internship helped you prepare for work after your graduation?
Would you like to work for this organization?
Would you like to work in a similar position somewhere else?

PROFILE 4.2

Mark Barnes, Eagleton Governor's Executive Fellow, 2010 and Ph.D. Candidate, Rutgers University (Piscataway, NJ)

Internships have been described as a "try before you buy" opportunity (Coco 2000), a chance to experience—and to gain experience from—an organization, position, job sector, or workplace without a long-term commitment. For Mark Barnes, a semester-long public sector internship played a pivotal role in helping him to decide on a future faculty career.

Mark received his bachelor's degree in geography and planning from West Chester University and his master's degree in urban studies from Temple University. He also gained valuable experience in the **applied aspects** of community development as an intern and employee with several organizations in Philadelphia. After enrolling in the Ph.D. program in geography at Rutgers University, Mark began considering a career in higher education but remained interested in urban issues and community-based organizations. On the advice of his dissertation adviser, he applied for the Eagleton Fellows Program as an "opportunity to step back and **gain another perspective** on politics." Rutgers students from any discipline are eligible for the fellowships, provided they have an interest in politics and government. Following a fall seminar, the fellows work part-time with a government agency or office during the spring semester.

Mark was writing his dissertation on the development of policies to address the impacts of snow events on urban transportation systems, particularly as climate change increases the vulnerability of cities to weather-related hazards. Based on his previous employment and his research interests, he was assigned to the New York/New Jersey Port Authority's Environmental Policy Division for his fellowship. While his duties focused more on climate change mitigation and carbon emissions reductions than on hazards, he met with policy makers and observed political processes at the state level, complementing his knowledge of municipal government and enhancing his political acumen. He also **gained firsthand knowledge** of state agencies, their objectives, and the challenges they face in funding transportation projects.

For prospective interns and fellows, Mark advises, "At the doctoral level, an intern has to **be mindful of fit** and be certain that the time is well-spent." He recommends that students seek placements that are relevant to their research topic. Whenever possible, he suggests that interns work with their colleagues to develop a project that connects to their academic interests and that they **identify a "champion"** within the organization who will serve as their mentor.

Looking back on his internship, Mark observes that perhaps the most valuable outcome was the realization that he wanted to pursue an academic career. Since returning to school, he had not done applied work for five years. His fellowship experience revealed that his "sensibilities have changed" during the time he spent as a Ph.D. student and adjunct instructor. Mark still describes community economic development as his passion and says he hasn't totally rejected the possibility of a career in government someday, but the return to the "9 to 5 daily grind" reminded him of why he decided to return to school in the first place: "I now know what I don't want to do!" he states. In Mark's case, "trying before buying" reaffirmed his love of teaching, the flexibility of the academic lifestyle, and his interactions with students, placing him squarely on the path to the tenure track.

—JOY ADAMS

consider when reflecting back on your experience. These questions might also serve as the basis of a final internship report to your internship coordinator. Basically, your report should cover: (1) your basic duties; (2) what you learned and gained from the experience; (3) any important or significant assignments that you handled; (4) a description of your work environment; (5) a discussion of your successes, as well as what and how you might have improved on your work; and (6) any final observations and comments.

FINAL THOUGHTS

Dr. Katharine Hansen (2010), creative director and associate publisher of *Quintessential Careers*, writes that the most valuable aspect of an internship experience is that it solves the "how-can-I-get-experience-if-I-have-no-experience" dilemma. Furthermore, in a 2008 survey by the National Association of Colleges and Employers (NACE)—a leading source of information about the employment of college graduates—95 percent of employers said that candidate experience is the most important factor in hiring decisions; nearly half preferred that the applicant's experience be generated from internships or co-op programs prior to graduation. Dr. Marilyn Macke, executive director of NACE, notes that hiring from intern programs has increased steadily since 2007. In a 2011 NACE *Internship and Co-op Survey* of 266 member employing organizations, Macke found that internships are an integral and ever-important part of the college recruiting scene and that employers expect to increase internship hiring by about another 7 percent over the year. Approximately 40 percent of their new college hires for 2011 were expected to come from internship and co-op programs, thus demonstrating the central role that experiential education plays in the overall college recruiting process (NACE 2011). Thus, an internship can be a win–win situation for you and your sponsor—you benefit by gaining valuable work experience as well as a deeper understanding of your major field of study, while your internship sponsor is able to observe your performance within their organization.

Once you have completed an internship and formed professional bonds with your supervisor and co-workers, stay in contact with them and other professionals through e-mail, contact at professional meetings, or online social networking sites, even after you have secured your first job. One never knows when professional contacts might be helpful for future job positions or collaborating on projects (Cross 2009).

Also, consider pursuing more than one internship, especially if your first experience did not meet your expectations. Use the internship experience to test-drive your career path. You may find that one niche in your specialization is a better fit than another. Alternatively, you might be interested in combining your knowledge of one geography subfield with another. For instance, if you are focusing on geospatial technologies, you might do an internship with a planning department of a major metropolitan area to gain experience using geospatial technologies for various urban planning projects.

Finally, if it seems to you that you face insurmountable obstacles to doing an internship, such as having to work part- or full-time to pay for your tuition and living expenses, find a way around these obstacles by talking to your school counselors, internship coordinators, professors, and family members so that you are able to take advantage of at least one internship opportunity before you graduate. Do some creative thinking about how to seize an opportunity—it's that important for your career.

REFERENCES

Baird, B. N. 2010. *Internships, practicum, and field placement handbook*, 6th ed. New York, NY: Prentice Hall.

Boehm, R. G., and S. Peters. 2008. *Careers/jobs in geography: Business cards of department graduates.* Texas State University-San Marcos: Department of Geography, Grosvenor Center for Geographic Education.

Bolles, R. N. 2010. What color is your parachute? A practical manual for job-hunters and career changers. Berkeley, CA: Ten Speed Press.

Brooks, K. 2009. You majored in what? Mapping your path from chaos to career. New York, NY: Viking Press.

Coco, M. 2000. Internships: A try before you buy arrangement. *S. A. M. Advanced Management Journal* 65, no. 2: 41-44. COS. College of the Sequoias, Visalia, CA. Internship website. http://cosinternship.net/ (last accessed 5 January 2011).

Cross, R. 2009. Grad expectations: The essential guide for all graduates entering the workforce. Penryn, Cornwall, England, U.K.: Ecademy Press Ltd.

Cross, R. L., and A. Parker. 2004. The hidden power of social networks: Understanding how work really gets done in organizations. Boston, MA: Harvard Business School Press.

Curran, S. J., and S. Greenwald. 2006. Smart moves for liberal arts grads: Finding a path to your perfect career. Berkeley, CA: Ten Speed Press.

Gralla, P., and J. Widman. 2008. Facebook vs. LinkedIn: Which is better for professional networking, job hunting, and collaboration? http://www.cio.com/article/193402/Facebook_vs._LinkedIn_Which_is_Better_for_Professional_Networking_Job_Hunting_and_Collaboration_ (last accessed 26 May 2011).

Hansen, K. 2010. College students: You simply must do an internship (Better yet: multiple internships)! *Quintessential Careers.* http://www.quintcareers.com/internship_importance.html (last accessed 24 May 2011).

Liang, J. 2005. Hello real world! *Student's approach to great internships, co-ops, and entry-level positions.* BookSurge, a DBA of On-Demand Publishing, LLC. www.booksurge.com.

National Association of Colleges and Employers (NACE). 2011. Internship and co-op survey: Research brief, 2011. http://www.naceweb.org/Research/Intern_Co-op/2011_Internship___Co-op_Survey__Research_Brief.aspx (last accessed 25 May 2011).

National Research Council (NRC). 2006. Beyond mapping: Meeting national needs through enhanced geographic information science. Washington, DC: National Academies Press.

Pollak, L. 2007. Getting from college to career: 90 things to do before joining the real world. New York, NY: HarperCollins Publishers.

Sabatino, C. 2011a. Online networking for professional and personal development. *Intern Coach.* http://interncoach.wordpress.com/2011/01/03/online-networking-for-professional-and-personal-development/ (last accessed 26 May 2011).

Sabatino, C. 2011b. Professional associations for professional development. *Intern Coach.* http://interncoach.wordpress.com/2011/01/03/professional-associations-for-professional-development/ (last accessed 26 May 2011).

SCA. Student Conservation Agency. 2011. *Conservation internships, student conservation association.* http://www.thesca.org/serve/internships (last accessed 8 January 2011).

The Princeton Review. 2005. The internship bible, 10th edition (career guides). www.RandomHouse.com.

UCGIS. The University Consortium for Geographic Information Science. 2006. *Geographic information science and technology body of knowledge.* Washington, DC: AAG.

Vogt, P. 2009. Which internship is best for you? *Monster College.* http://college.monster.com/education/articles/68-which-internship-is-best-for-you?page=1

Wise, C. C., S. Hamadeh, and M. Oldman. 2009. *The Vault guide to top internships, 2009 edition.* New York, NY: Vault, Inc., Vault Career Library.

5

Professional Networking

Tina Cary

Professional networking is the process of developing relationships with people in the field of one's occupation. That is, the first—and often the strongest—bond linking you and the people in your professional network is your sharing of one or more of these things: education, training, career experience, or intellectual interests. Whether you have similar or different views on politics and whether you root for the same or opposing sports teams may not even come up for a while; knowing the same professors and courses, working in similar organizations, or using the same geospatial technologies can provide the basis for many engaging and mutually beneficial conversations.

The words "mutually beneficial" bring up another element of professional networking: people who focus on developing relationships only to get what they want rarely find long-term success. In contrast, people who know that giving first will later result in receiving a benefit understand a key aspect of the networking process. If you look at the people whose career accomplishments you most admire, you are likely to find they have strong relationships with a variety of other people in the same field—people in different career stages, different geographic areas, different specialties, and (or) different kinds of employment. The strength of such relationships is more important than their number.

Because a professional network is a kind of social connection, many of the guidelines for developing, sustaining, and enlarging one's professional networks are familiar. The process of developing and sustaining relationships has the same essential elements, whether the focus is professional or social: listening attentively, sharing yourself and your knowledge, asking for what you need, and reciprocating are integral parts of the process. In addition to thinking of networking as the process of developing relationships, one can look at it as the development of social capital.

> Social capital is the accumulation of resources developed in the course of social interactions, especially through personal and professional networks. These resources include ideas, knowledge, information, opportunities, contacts, and, of course, referrals. They also include trust, confidence, friendship, good deeds, and goodwill. (Misner, Alexander, and Hilliard 2009, locations 206-209)

Invest Your Time

How much time does professional networking take? Perhaps a better question is, "How much time does professional networking save?" Let us first look at how much time it takes; the later sections on the benefits of networking will include discussion of the time-saving aspect. Professional networking is definitely an important activity, though many people rarely consider it urgent (Foote 2009). Elaborating on the title of Harvey Mackay's (1999) book, *Dig Your Well Before You're Thirsty: The Only Networking Book You'll Ever Need,* if you wait until it is urgent, you may not have the energy to do it. How much time should be given to networking will vary with your goals, priorities, and responsibilities. For example, geographers who are employed in sales or marketing positions—whether selling satellite data or image-processing software or other geo-related products or services—find that they spend a lot of time on activities that fit the definition of professional networking: meeting people, exchanging business cards, asking them about themselves and their challenges, identifying mutual acquaintances, offering help or ideas, evaluating opportunities for mutual benefit, and planning a next meeting. Best-selling author and networking consultant Ivan Misner recommends that salespeople spend more than half their time in networking activities (Misner and Donovan 2008, location 608).

If you are seeking a new position, you will spend more time on such activities than if you are currently in a position that is right for you (and is relatively secure). Remember that even in a good situation, developing and maintaining your network will help you not only with your current job responsibilities, as discussed below in the section "Benefits to You," but also when you seek a promotion or transfer.

In general, a modest amount of time spent daily and weekly will be manageable and yield results. Having goals and a written plan to follow will help you maximize the return on your time. Surveys of businesspeople have found that 52 percent spend 4 hours per week or less on professional networking, while 27 percent spend 8 hours or more per week (Misner and Donovan 2008, location 602). If you are just getting started, I encourage you to schedule an amount of time that is modest enough to allow you to be very strict about keeping the commitment to yourself. For example, in 15 to 30 minutes per day, you can review the names of people you already know and make a phone call or send an e-mail to someone you have not talked to for a while to renew the connection, share or request information, ask for help, or extend a compliment or an invitation. As you begin to see results, you will want to increase the time while maintaining the firm commitment.

You probably already have a routine for handling e-mail, or at least you are well aware that you need to have one. Responding to e-mails that you receive from individuals you know

SIDEBAR 5.1

Make Your Networking Plan

Identify what you want networking to do for you. Be specific: what one professional goal can you set and achieve in a specified time period (the next quarter? the next year?) by networking?

Identify what you will deliver as benefits to your organization. Why is your company, your agency, or your nonprofit better off if you do this networking than if you don't?

Identify tactics and the professional networking tools you will use to reach the goals you identified above. What benefits will the geospatial community realize as a result of the networking activities you will be doing to meet your goals?

gives you an opportunity to enhance an existing relationship by the timeliness and usefulness of your reply. If you have a routine that's working for you, consider using it as a starting point for developing, expanding, and fine-tuning your networking activities.

When should you begin to develop your professional network? Today! Whether you are in college, your first job, or midcareer, you are surrounded by people with whom you share experiences every day. Start with people you respect and trust. They all know people and information you do not, and they may be interested in the people and information you know. Be open to the idea that opportunity is all around. One such example occurred a few years ago for a couple who owned an aerial photography firm. They wanted to sell their company and retire. The woman went to the fabric store where she regularly purchased quilting supplies. At checkout, the clerk asked her how business was going, and she responded that they were ready to sell the company. The person behind her in line said, "My husband and I are looking for a business to buy. What kind of company is it?" From that beginning, a business agreement was reached and the company sold.

Another example of an unexpected opportunity came to me while I was in my first position after graduate school. It was a faculty position, and I was so stressed that I had to make a change. I called a man I had worked for before graduate school. I told him my situation and asked his view of whether the stress might be unique to that particular department or whether I would be better suited to a business or government environment. He had been employed in academia and government before moving to the position in industry he occupied when I called. In combination with the fact that he knew my skills and work habits, I thought his advice would be particularly helpful. He answered my questions and then said, "You know, this is interesting timing. We need a training coordinator here, and I haven't yet seen any applications from people with appropriate credentials and experience." I sent my résumé, interviewed, and was hired. Joy Adams highlights those sorts of transitions in Chapter 3 on "Switching Sectors."

How to Begin

How do you build a professional network? When I started graduate school in the Geography Department at Columbia University, my adviser said, "There are two professional organizations you should join: the Association of American Geographers (AAG) and the American Society of Photogrammetry [now the American Society for Photogrammetry and Remote Sensing (ASPRS)]." I took his advice, and it worked well for me, so I pass along this version to you: Identify two professional organizations that align with your interests and become an active member. To the question, "How do I choose?" my answer is a generic version of the advice I received: One should relate to your discipline of study and the other either to applying your discipline's principles to the same kinds of problems you face in your work or to using the tools you are most interested in using.

In addition to attending professional meetings in a discipline (AAG, for example), you may want to network with professionals who share other interests, as is noted in the chapters on business, government, and nonprofit organizations in the next section of this book. That is, geographers in business may find opportunities for networking in professional organizations in marketing, logistics, and supply chain management, or site location and real estate, and so on. Government employees may want to network with others in areas such as urban and regional planning, environmental protection, property assessment, or other government functions. Geographers working for nonprofits may seek organizations that focus on mapping if, for example, they map environments important to the organization's mission (such as wetlands for Ducks Unlimited), or they may want to network with peers engaged in target market segmentation if their responsibility is to analyze the socioeconomic characteristics of repeat donors in order to market to other people with similar characteristics. If you use specialized technology, then attending a specialty conference may also be appropriate. Examples of specialty conferences include the International Lidar Mapping Forum, which serves professionals interested in terrestrial lidar, and user conferences sponsored by software and hardware companies for their customers.

If your question is not "How do I choose?" but rather "How do I know what my choices are?" you will need to investigate what professional societies exist for your areas of interest. On the Internet you can look up the biographical data of professionals you admire (professors, authors, people in your government agency, the head of your nonprofit) to learn what organizations they mention. If you are at the beginning of your career, you can turn to your academic adviser for recommendations. If you are employed, ask whether your organization reimburses any particular memberships; however, if you wish to create opportunities to leave your present place of employment, this information may be of little or no importance. If you are midcareer and seeking to change from one sector to another, find someone in your network who has experience in the sector of interest and ask what they would suggest and why. Ultimately, you will have to determine what best matches your own professional interests and goals, and make the final decision yourself. As your goals and interests change, remember to evaluate whether your memberships are serving you well, or whether you may need to make new choices.

Earlier I recommended becoming an *active* member of two organizations. You will benefit much more from your membership if you do more than read the journal that all members typically receive from a professional organization. As Ken Turnbull points out in his profile, volunteering is not optional; it's necessary for career advancement and personal growth. If the organization has Specialty Groups or Special Interest Groups by whatever name, choose one in which you can make a contribution. I encourage you to volunteer for something specific, but only if you are willing to commit time and energy. Volunteering for something and not doing it or doing it poorly will not help you professionally; people will notice and remember commitments not kept as surely as they will remember contributions.

Such groups often need people to organize sessions for an annual or regional meeting, people to present papers in those sessions, someone to manage a website, contributors and an editor for a newsletter, or people to work on a special project. When you choose a way to contribute, you have a reason to get in touch with people and are generally given the contact information to do so. By volunteering, you are taking action that proves you care about the field and that you want to share your capabilities and knowledge to help others. This is all part of a very important component of professional networking: By making a contribution, you trigger reciprocity, so that people will be inclined to help you when you need something.

BENEFITS OF PROFESSIONAL NETWORKING

Professional networking can yield many benefits as a career development tool. Some benefits come directly to you; other benefits may come first to your employer and, because they increase your value to your organization, benefit you indirectly. Yet other benefits accrue to your professional community first and, again, benefit you indirectly. The following subsections discuss each of these three categories of beneficiaries.

PROFILE 5.1

Ken Turnbull, Accredited Land Consultant (Denver, Colorado)

With an estimated 4,000 active professional contacts, Ken Turnbull knows a thing or two about networking. His career path has traversed nearly four decades, leading him to a variety of positions related to the geosciences and geospatial technology. Whether generating maps and models to support petroleum exploration, managing sales and contracts within the aerospace industry, advising clients on land, mineral, and water rights, or taking a seven-year hiatus to raise ostriches, Ken has observed time and time again that who you know is just as important as what you know.

Ken regards networking as an essential aspect of career development. "The sooner you begin expanding your network, reaching out to others, and maintaining a professional social life, the better off you will be," he advises. He points out that employers hire in order to solve problems. Because jobs are therefore inherently about meeting the needs of others, **networking is critical**. Your professional contacts can help you to understand a specific industry or company's needs so you can make a compelling case for why you're the right candidate for a job opening.

"We human beings like to do something familiar rather than something fearful and unknown," Ken observes, noting that most employers prefer to hire personal acquaintances or people within their employees' networks. While new technologies have made virtual networking ubiquitous, he thinks online communication should be considered "a primer" for **face-to-face networking**; it is best utilized to make basic introductions in advance of an in-person meeting whenever possible. "LinkedIn is the only thing I've found that I like," he states, adding that the "Groups" feature is especially valuable for making connections with professionals who share your interests.

In the digital age, Ken feels that personal contact is more important than ever. He recommends involvement with **professional societies and organizations** as an important first step to developing your network while supporting the vitality of your chosen field. "Volunteering is not an option; it is an absolute requirement for career advancement and for personal growth," he explains. "Individuals giving back to our geospatial community are what keeps industries sound and enable the cross-pollination of ideas between related disciplines."

One of Ken's contributions is his stewardship of the "Rocky Rogues," a group of Denver-area geospatial professionals who gather twice each quarter. From just a handful of Ken's co-workers from his previous position at Digital Globe, the group has grown to 500 subscribers. Ken fronts several hundred dollars for the meeting space and pizzas, requesting a modest $5 donation from attendees to offset a portion of the costs. While he continues to underwrite many of the events, the Rogues success has recently led to occasional corporate sponsorship. Along with networking, participants are given opportunities to hone the **social and communication skills** needed in most jobs. "Many technical people in particular tend to shy away from social events until they get comfortable and then of course, they look forward to it," Ken observes. "If you are not a natural at socializing, force yourself to participate until it becomes familiar and comfortable—you will enjoy it." Whether attending formal events or informal gatherings, Ken strongly advises that even students come prepared with business cards and hand them out freely to new acquaintances. While amassing thousands of professional contacts reflects concerted effort over a long and diverse career, it's never too late or too soon to start establishing relationships with professionals in your field. As Ken observes, "People are important, even those you do not think are related directly to your career needs at this time."

—JOY ADAMS

Benefits to You

Networking provides many benefits, including improving your job performance, increasing your value to your organization, enhancing your reputation, and maximizing your career opportunities.

JOB PERFORMANCE By connecting with individuals who are facing the same challenges as you, or who faced them recently, you can save time and avoid making mistakes that other people have made. Here are some specific examples to convey the idea that networking can be useful in different sectors and in jobs having very different responsibilities:

- You may learn from members of your network about a new software package for project planning.
- You may get good advice about the pros and cons of a particular mobile device from someone who already uses such equipment.
- A member of your network may be able to recommend a great online resource for developing a geospatial privacy policy for your organization.
- Someone may be able to recommend a strategy that your nonprofit organization can use to segment potential donors and improve the results of fundraising efforts.

In those examples, you will get better results faster than if you do not get advice from your professional network. You can think of your network as your council of advisers. If you choose your advisers wisely, their interests and yours will be complementary, and you will be able to find many opportunities to work together for mutual benefit.

VALUE TO YOUR ORGANIZATION By using the advice of your network to get better results faster, you will increase your value to your organization—your value today compared to your value yesterday and your value compared to the value of other people in the organization.

ENHANCED REPUTATION Of course, the relationships that comprise your professional network will be sustained over the long term only if you reciprocate. And it is in the reciprocation that you enhance your reputation. An important point to realize is that your reputation is a combination of factors, only one of which is your technical competence. In order to be the "go to" person, you need technical competence, of course, but that is not enough. Others have similar competences; if you are easy to work with and you keep your commitments, your network will know, and they will spread the word on your behalf by mentioning you in favorable ways.

CAREER OPPORTUNITIES As you become known for being competent, easy to work with, keeping your commitments, and returning favors others do for you, career opportunities will become available to you that will be unavailable to others with similar competence who have not developed a good professional network.

Benefits to Your Organization

Your networking activities can provide your organization interconnected benefits because they can save time and money as well as reduce risk.

SAVE TIME AND MONEY When your networking saves you time and money, as discussed above under "Job Performance," those savings benefit your organization. A mapping company whose vice president of marketing has a broad and mature network received double the value from her marketing budget, compared to what the money could have accomplished for someone without such a network. She knew the advertising representatives for the industry publications relevant to her company's target markets and she described her situation and interests to them, so they knew to call her if a publication deadline was near and space had not sold, offering a significant discount.

A sales executive in the geospatial arena has a broad and mature network. She knows the goals and challenges of the people in her network and the kinds of products or services they need. This is invaluable information as she moves from one company to another. Her knowledge and connections enable her to focus on solving problems; she does not use her valuable time making cold calls. She has been able to meet sales goals for a variety of organizations, delivering revenue and market share to her employers.

REDUCE RISK Another benefit to your organization can be in minimizing the risk associated with adopting new technology. If you are a member of a group that focuses on a particular software package and shares experiences, you can receive guidance about upgrades. Group members will identify any problems and fixes they have encountered, so your organization will have minimum downtime when it does the upgrade. Your organization also benefits from your network in the staffing process. When positions become open, you can make the opportunities known to your network; while you are helping those people (and they may potentially reciprocate), you are also increasing the pool of applicants from which your company gets to select. You can also talk to the hiring manager or the human resources department about how well the candidates you know would fit the culture of your organization, something of great value that is generally hard to learn in any other way. Your professional networking can deliver benefits to your organization in helping it hire the best person and minimize the risk of a costly hiring mistake.

Benefits to the Larger Community

Professional networking also delivers benefits to your professional community at large. This is another way of looking at the service you contribute as part of building your network. For example, you may identify one or more of the following volunteer activities as offering you opportunities to meet and work with professionals you want to know: organize a conference session, manage a website, make a presentation, write a letter of recommendation, or review an article. While your primary motivation for volunteering may be to become known to specific individuals, your actions will also provide benefits to the broader professional community. For example, the conference session you organize will enable people to share their research and help audience members to learn about new developments. The website you manage will provide a resource for people to know about the activities of the group and will give employers and employees a place to find each other. Your presentation will enable other people to know about your research and learn where the research frontiers are in your specialty. The letter of recommendation you write will help an organization to find a good employee and a colleague to find a good position. All of these results provide benefits to your community while enhancing your standing within it.

TOOLS FOR NETWORKING

The tools available to facilitate professional networking fit into three categories: (1) professional meetings; (2) other traditional tools; and (3) social media.

Professional Meetings

The classic tool for networking is the professional meeting. Professional organizations serve many different needs for individuals, offering opportunities to find like-minded people, to expand your network, to become informed about research frontiers and new products and services, to share research results, to find employees or employment, to gain speaking experience, and to reconnect with people you have not seen for a while. The organizations use conferences and publications as tools for meeting these different needs.

Geographers have been developing relationships with colleagues for more than a century, as indicated by the longevity of some of our professional organizations—the Royal Geographical Society was founded in 1830, the American Geographical Society in 1851, and the Association of American Geographers in 1904—long before "network" became a verb (Cary and Associates 2011). At the SPAR International 2011 conference on three-dimensional imaging for design, construction, manufacturing, and security planning, at least three new organizations were proposed (Schutzberg 2011). To me, this says that people still want to look someone in the eye and share meals and handshakes. Although the Internet or other contact methods can be beneficial, they supplement rather than replace face-to-face meetings for developing one's own advisory council and accumulating social capital.

How much benefit you, your organization, and the community derive from your participation in any particular meeting will be a function of how strategically you think about the opportunity, how well you plan for it and execute your plan, how flexible you can be when faced with a tremendous unplanned opportunity, and how well you follow up. How you participate in a professional meeting will evolve through your career and will be shaped by your goals and your organization's goals for your participation. If you are in an early stage of your career, your plan could include checking the program for presentations on topics of interest, identifying (at least) two presenters per day that you would like to meet and attending their presentations, identifying a few topics of conversation for any social events, and confirming you have a good supply of business cards.

Further along in your career, preparation may start much earlier, with identifying a topic about which you would like to make a presentation, submitting an abstract, making plans to get together with people already in your network, and coordinating with colleagues where you work to make a list of questions to ask vendors in the exhibit hall about products or services your organization is considering purchasing. If you are unemployed, you will want to check the program to see if a job fair or other professional development activity will be available. You will also want to register for any opportunities that support your goals, as well as update your professional website (if you have one) and make several copies of your résumé. It would also be a good idea to study the meeting program to see if organizations you are interested in are presenting or exhibiting, and create a personal schedule to meet with them. At a later career stage, you may be involved in planning the program, attending committee or board meetings, making a keynote address, and participating in other activities that contribute to the community while sustaining and enhancing your reputation and your professional network.

Follow-up often seems to be the weakest link in networking at professional meetings. I encourage you to make notes on the business cards you receive to remind yourself later what you and the new contact have in common, and to identify anything you promised to send them. Either in the evenings at the conference, on the trip home, or as soon as you get to the office, review the cards and allocate the time you will need to send a quick note and any promised information (preferably in a week or less), and enter the contacts into whatever system you use, along with notes about where and when you met the person and what you have in common. If appropriate, also enter into your schedule when you will call or e-mail, or any other plans for future contacts. Too often people return to the office, look at business cards, remember little about the person, and throw the cards away. This effectively converts the valuable time you spent with new people into wasted time—an unnecessary forfeit when making notes on the business cards is so easy.

Other Traditional Tools

Personal visits, the telephone, and postal service come readily to mind as traditional tools. E-mail and listservs are also considered traditional tools since they have been around for more than a decade. In fact, because e-mail has come to replace many communications formerly done by mail, sending a handwritten note when you want to thank or congratulate someone is now widely recommended as a way to make a positive impression. For the nearby people in your network, a personal visit is an excellent option. By scheduling one meeting per month, whether for a meal or a drink or an event, you will deepen your relationships with a dozen people a year. Calling people remains an essential tool. Hearing a familiar voice and sharing a few minutes on the telephone is an enjoyable way to accomplish some objectives. E-mail allows you to schedule your time for networking to fit your life and time zone, and it gives you a way to let someone know you are thinking of them and why.

For readers not familiar with listservs, a description of and link to a particular example may help convey their utility. GIS Colorado (http://www.giscolorado.com/) operates a listserv where members routinely seek information and advice on a variety of topics ranging from where to find infrared photography for a particular county to what experiences people have had with a particular plotter to what online geographic information systems (GIS) courses people have found useful. Over time the people who regularly provide useful information become known to members of this list and beyond for their expertise. This list is so useful that people from Pennsylvania, Texas, and other locations outside Colorado are members.

Social Media

Social media is the term for a variety of tools that enable sharing and collaborating with a high degree of interactivity over the Internet and mobile devices. Rather than trying to cover the use of every kind of social media for professional networking, this section provides brief descriptions of some Internet-based tools for sharing, collaborating, and interacting with others. Though there are many more, I have picked a few examples that I have found particularly useful in my own work to show how people in the geo-arena are using social media (Table 1). Glenn Letham's profile conveys how he uses social media to build his business and stay informed about the industry. In order to choose the right tools for you, start by reviewing your goals, including what people you want to connect to and why, and what you want to give to the geo-community.

PROFILE 5.2

Glenn Letham, Co-founder, Spatial Media LLC, and Managing Editor, GISuser.com (Vancouver, British Columbia)

Glenn Letham might be a jack-of-all-trades, but don't call him a master of none. The self-described geographer, GIS professional, and "technology evangelist" is the co-founder of four popular geographic and mobile technology Web magazines. He also regularly blogs, participates on technology panels, and has attended over 100 major tech conferences and events around the world. In his free time, Glenn is an avid user of social media. Needless to say, he knows a thing or two about technology.

A native of British Columbia, Glenn is a graduate of the University of Victoria. He started out as a business major, but a few geography professors "planted the seed" that would grow into a new area of interest. "They were my biggest influences," he explains. "I found that geography really resonated with my interests in computers and technology." After working as a GIS analyst at a consulting firm for several years during the 1990s, Glenn's job became a casualty of the "dot bomb era." With free time on his hands, he decided to explore the blogosphere: "I like to get my hands dirty, so I started my own blog—*Spatial News*—and it just kind of took off." Soon after, Glenn co-founded Spatial Media with his partner Allen Cheves, and in 2004, the pair launched GISuser.com. The website provides the latest developments, analysis, and reports concerning GIS and geospatial technologies. According to Glenn, "Everything we [publish] has some kind of spatial component."

Because Glenn must stay abreast of the latest technology, he closely monitors the "pulse of the industry" through his expansive network of professional contacts. While it helps to have years of experience in the field, he uses **social media outlets** such as LinkedIn, Facebook, and Twitter to grow his network: "I want my name and my face to be recognizable. People often feel they know me even if they haven't met me." Glenn's approach has opened many doors. Recently, he has begun using social media as a means of finding or organizing local events for geography and GIS professionals. Examples include Ignite Spatial, an "un-conference" held in conjunction with the GIS in the Rockies meeting to provide a social activity for attendees, and "YYJ Geo Geeks," a tech-savvy group of local academics, government workers, and consultants who meet for drinks each month. (YYJ is Victoria's airport code.) According to Glenn, "People are really receptive to these events, and social media definitely helps with recruiting speakers and getting turnout."

For anyone on the job market, Glenn says that LinkedIn is "the number-one resource I'd suggest" for networking and finding opportunities: "Even if you're not a 'pro' user, the quality of connections and groups is top-notch." He also recommends using Twitter as a means of "**branding yourself**," but he notes the importance of distinguishing between personal and professional communications: "Until you build your network and become established, you need to be very conscious of how you portray yourself online." Glenn's experience demonstrates that networking can be useful for more than just finding jobs. Perhaps as importantly, it can provide a means of honing the **interpersonal skills** often overlooked in formal education: "Social skills are very important in business, even for 'techie' or programmer types because they need to be able to communicate to others what they're doing."

—**MARK REVELL**

In the category of multimedia sharing, SlideShare is used to share slide presentations. Aerial Services, LiDAR_MN, and blogger James Fee have posted several presentations there. If you have presentations that convey your expertise, you can post them on SlideShare, and then, as part of your professional networking activities, you can send the link via e-mail to colleagues you think would be interested. YouTube is used to share short videos (limit 15 minutes). The Carbon Project, the U.S. Geological Survey (USGS), and ASPRS are some of the business, government, and nonprofit organizations that use YouTube. Peter Batty uses Vimeo to share longer videos of his presentations. Presentations can be shared live over the Internet and preserved for later viewing. Ignite Spatial Northern Colorado used Ustream for the event in May 2010 and Livestream for the event in September 2010. Glenn Letham of GISuser.com uses Flickr to share photographs, typically from geo-conferences he attends. The trade magazine *GIS Development* uses Picasa to share albums of photos from events. If you like to take pictures at events you attend, posting them can remind people of you as well as the event. VerySpatial.com uses podcasts to share a weekly episode about geography and geospatial technologies. This activity gives the hosts a reason to contact anyone they want to interview and is a great way to combine giving to the geospatial community and networking.

Collaboration tools include wikis; a wiki is a website to which many users contribute. OpenStreetMap and Wikimapia are map examples of wikis, and wiki.GIS.com is a text-based geo-example. Many people update OpenStreetMap for their local area as a weekend hobby. If you want to join the open data community, contributing to OpenStreetMap will give you opportunities to get to know people in that community. After the 2010 earthquake in Haiti, volunteers updated OpenStreetMap data and by doing so contributed to the rescue effort.

Bookmarking tools also facilitate collaboration. You can bookmark any URL you find interesting and give it tags that are meaningful to you. The Open Geospatial Consortium encourages members to use Delicious to bookmark articles related to open geospatial standards. If you are interested in such standards and want to network in that community, consider using Delicious to bookmark sites and tag them to help yourself and others find them.

Blogging is a form of social media used to share information and ideas, and to create and enhance one's reputation for being well informed and offering ideas worthy of attention. Examples include Geoff Zeiss's "Between the Poles," Gene Roe's "LiDAR News," and Gretchen Peterson's "GIS Cartography." Microblogging (messages limited to 140 characters) is done on Twitter, Tumblr, and other platforms. In Table 1, you will find URLs to three lists of Twitter users: geospatial companies, U.S. federal agencies involved in mapping, and state geographic coordinators.

Social networks are another category of social media used to build communities of people with common interests. Some networks are primarily social, whereas others have professional networking as their focus. LinkedIn is the leading network for professionals. In March 2011, LinkedIn announced that it had 100 million members, 44 million in the United States and 56 million outside the United States (Weiner 2011). LinkedIn offers many tools (Elad 2008) for connecting with like-minded professionals—profiles, groups, recommendations, jobs, company information, and LinkedIn Answers (post and answer questions). In his profile, Glenn Letham identifies LinkedIn as the best resource for anyone on the job market.

As a member of LinkedIn, you can create a profile with your education and work history, and choose whether to have the profile available to anyone. If you do create a public profile, you can then use a link to that page in your e-mail's signature block or on your résumé to invite people to connect with you. You have access to your account from anywhere, allowing you to access the contact information for everyone to whom you are connected—a convenience for reaching people while you are on the go. In his profile, Ken Turnbull notes that the "Groups" feature is very valuable for finding people with shared interests. For example, choose Groups from the search box menu, enter "geospatial" in the search box, and you will find that numerous groups already exist (172 on 29 April 2011, up from 136 on 15 September 2010), created by professional organizations, event planners, companies, and individuals. Looking through the descriptions will help you determine which groups are most likely to help you advance your goals. Groups can be open or closed; only members can see closed groups, while open groups can be seen by anyone. Table 5.1 includes links to four geospatial groups that are open.

If you do not find a LinkedIn group suitable for your goals, you have the option to create a group that can become a community and facilitate discussion. Some groups on LinkedIn are not well moderated, and unrelated material gets posted and left for days. I encourage you to monitor at least half a dozen groups for a month to see how they work and figure out if you can accomplish the most by contributing to a few key existing groups, or whether creating a new group and moderating it is necessary in order to meet your goals. Useful groups share these characteristics: new material gets posted on a variety of interesting topics, typically a few times a week, and most posts start discussions. In my opinion, the moderator of a group holds the key to the value of the group. By responding to the posts of others, removing inappropriate posts, and asking questions of the group, the moderator facilitates the development of a community.

Facebook is generally considered a personal social network; however, Facebook applications ("apps") and pages are being used by business, government, and nonprofit organizations such as Trimble Outdoors, the USGS, and the Geospatial Revolution Project (see Table 5.1 for URLs).

TABLE 5.1 Some Examples of Social Media Useful for Professional Networking

Slide Presentations	
LiDAR_MN	http://www.slideshare.net/LiDAR_MN
Aerial Services	http://www.slideshare.net/aerialservices/
James Fee	http://www.slideshare.net/cageyjames
Short Videos (limit 15 minutes)	
The Carbon Project	http://www.youtube.com/thecarbonproject
USGS	http://www.youtube.com/USGS
ASPRS	http://www.youtube.com/ASPRS
Longer Videos	
Peter Batty	http://www.vimeo.com/pmbatty
Photos	
Glenn Letham	http://www.flickr.com/photos/gisuser/
GIS Development	http://picasaweb.google.com/GISDevelopment.net
Streaming Video	
Ignite Spatial Northern Colorado—May 2010	http://www.ustream.tv/recorded/6755716
Ignite Spatial Northern Colorado—Sept 2010	http://www.livestream.com/ignitespatialnoco
Podcasts	
A Very Spatial Podcast	http://veryspatial.com/avsp/

Wikis	
OpenSreetMap	http://www.openstreetmap.org/
Wikimapia	http://www.wikimapia.org
wiki.GIS.com	http://wiki.gis.com
Bookmarking	
Open Geospatial Consortium	http://www.delicious.com/search?p=Open+Geospatial+Consortium
Blogs	
Geoff Zeiss	http://geospatial.blogs.com/
Gene Roe	http://blog.lidarnews.com/
Gretchen Peterson	http://www.gretchenpeterson.com/blog/
Microblogs	
Geospatial companies	http://twitter.com/tinacary/geo-companies/members
Federal agencies involved in mapping	http://twitter.com/tinacary/federal-twitter-accounts/members
State GIS Coordinators	http://twitter.com/nsgic/state-gis-coordinators/members
Social Networks—LinkedIn	
Colorado GIS / Geo Techology Professionals	http://www.linkedin.com/groups/Colorado-GIS-Geo-Techology-Professionals-137815
National Geospatial-Intelligence Agency	http://www.linkedin.com/groups/National-GeospatialIntelligence-Agency-1820567
GIS Training and Education	http://www.linkedin.com/groups/GIS-Training-Education-1845098
GIS Pros For Hire	http://www.linkedin.com/groups/GIS-Pros-Hire-3735632
Social Networks—Facebook	
Trimble Outdoors	Fan page: http://www.facebook.com/trimbleoutdoors App: http://apps.facebook.com/trimbleoutdoors/
U.S. Geological Survey (USGS)	http://www.facebook.com/USGeologicalSurvey
Geospatial Revolution Project	http://www.facebook.com/geospatialrev

Social media tools in use by the Association of American Geographers exemplify the range of tools professionals find useful (see Table 5.2 for URLs): a knowledge community, a facebook page, a twitter account with hashtags designated for meetings, a LinkedIn group, and a video series on blip.tv.

TABLE 5.2 Social media available from the Association of American Geographers

Knowledge Communities	http://community.aag.org
Facebook	www.facebook.com/geographers
Twitter	www.twitter.com/theAAG, @theAAG, #AAG2011
LinkedIn	www.linkedin.com/groups/Association-American-Geographers-53689
BlipTV	http://aag.blip.tv

IN CONCLUSION

Professional networking is an important component of managing one's career and must be given priority and regular attention, even though it seldom seems urgent. This chapter provides guidelines for networking, with a focus on people working in, or seeking work in, business, government, and nonprofit organizations. The benefits of networking are described in three categories: those that accrue to the individuals, to their organizations, and to the larger community of geoprofessionals. Professional meetings and other traditional tools are now supplemented by social media as ways to sustain and enhance one's network.

REFERENCES

Cary and Associates. 2011. Professional networking. *News and Notes from Cary and Associates* 19 (21 April). http://community.icontact.com/p/caryandassociates/newsletters/news/posts/21-apr-2011-keepup-to-date-with-cary-and-associates (last accessed 29 April 2011).

Elad, J. 2008. *LinkedIn for dummies.* New York, NY: Wiley.

Foote, K. 2009. Time management. In *Aspiring academics: A resource book for graduate students and early career faculty,* eds. M. Solem, K. Foote, and J. Monk, 5–15. Upper Saddle River, NJ: Prentice Hall.

Mackay, H. 1999. *Dig your well before you're thirsty: The only networking book you'll ever need.* New York, NY: Crown Business.

Misner, I. R., D. Alexander, and B. Hilliard. 2009. *Networking like a pro: Turning contacts into connections.* Irvine, CA: Entrepreneur Press. Kindle edition.

Misner, I. R., and M. R. Donovan. 2008. *The 29% solution: 52 weekly networking success strategies.* Austin, TX: Greenleaf Book Group. Kindle edition.

Schutzberg, A. 2011. New and proposed survey related organizations. *Directions Magazine*, All Points Blog (8 April) http://apb.directionsmag.com/entry/new-and-proposed-survey-related-organizations/173706 (last accessed 29 April 2011).

Weiner, J. 2011. 100 million members and counting... LinkedIn Blog (22 March) http://blog.linkedin.com/2011/03/22/linkedin-100-million/ (last accessed 30 April 2011).

6

Geography Careers in State and Local Government

William M. Bass and Richard D. Quodomine

Congratulations! You've got your cap, your gown, and you've moved your tassel. You are a newly minted geographer, ready to analyze the landforms, economy, or demographics of the world. Or perhaps you've heard the buzz about geographic information systems (GIS) and have decided to pursue a GIS certificate or graduate degree in preparation for a career in the geospatial technology industry. In any case, you may find it difficult initially to identify an opportunity for yourself in government, with a corporation, or in the nonprofit sector. As you scan job listings, you come to a sinking realization: very few say "Hiring Geographers. Please Apply." But don't let this frustrate you. Both of us reached this point some time ago, but we just as soon figured out that just because many employers don't advertise for "geographers," it doesn't mean that they don't need geographic skills.

In the public sector, your challenge will be to seek opportunities for applying your geographic abilities to address the needs of government agencies. As noted in a recent issue of the U.S. Department of Labor's *Occupational Outlook Quarterly* (Crosby 2005, 4):

> In the last decade, the federal government has launched several new global positioning satellites. These, together with some commercial satellites, greatly increase the accuracy and amount of geographic data available. At the same time, new Geographic Information System (GIS) software can process those data with greater speed and flexibility. This technology creates new career possibilities for people who understand geography and who can process and use geographic information.

Since very few state and local governments have a Geography Department in their organizational hierarchy, this chapter will focus on the various types of work performed by geographers across a wide range of agencies (BLS 2010a). Geography professionals practice in a variety of organizational environments at the state and local level, and work on interdisciplinary teams that are comprised of professionals from backgrounds other than geography. Thus, practicing geography in this type of environment lends itself to professionals with interdisciplinary skills and a strong interest in working with others whose skills are different from their own. This chapter will also provide some resources that you may find useful in your job search, and will conclude with some practical advice on professional networking.

THE NEED FOR GEOGRAPHERS IN STATE AND LOCAL GOVERNMENT

Why do people choose to practice geography in government at the state and local level? Most who do so are not seeking to become the next overnight millionaire or world-famous celebrity. Instead, people choose to work in government because they feel it allows them to make a difference in their communities. Another reason many choose a career in government is that it offers a healthy work-life balance, with little overnight or extended travel requirements. You probably won't find yourself on a plane every Monday morning leaving your family for an entire week, then getting back late on Fridays, only to repeat the process for weeks on end.

Furthermore, many government jobs focus on longer-term contracts and project cycles than some private sector jobs, and thus offer a bit more continuity and stability. We are all aware of recent spending cuts at all levels of government, so the assumption that all jobs in government are totally secure is not perfectly accurate. However, organizational structures in government tend to be more stable than those in the private sector, as reorganizations may require legislative mandates rather than only the desire of corporate leadership to restructure or merge the company.

Finally, positions in local and state government offer very challenging and rewarding opportunities that can't be matched in private industry. Seeing a long-term community development project reach completion, or working hard to improve the social, educational, or transportation services in your community, region, or state, are just a few examples. Tapping into what you are passionate about and targeting an area that offers unique and exciting opportunities in your field of interest are the best ways to start down the path toward a rewarding career.

Geographers play an important role across many functional areas of state and local government. They have the advantage of being able to understand geographic areas by virtue of their extensive education in exploring interactions and linkages within and between places and regions. Many government organizations require a GIS expert, a specialist in statistics, or a person skilled in databases. In addition, in these times of retrenchment and a general decline in civil servant numbers (BLS 2009), agencies may not have the freedom to hire all the specialists they need, so public employees often have to address challenges that are new and might have been handled previously by a colleague. This is often where geographers can prove their skills to their employers and the taxpaying public.

Delivering critical information through geographic visualizations is another way geographers are contributing to the work of state and local governments. Maps can often tell a story better than tables and figures, and telling a story requires a map to convey both expertise and simplicity.

When communicating with professional colleagues or experts in the field, a government employee can feel free to utilize more technical or detailed facts and figures. However, a precise and technical subject does not require verbosity for effective communication. In fact, the opposite is usually true. Officials need advice and, because they are often facing tight deadlines, this advice needs to be communicated succinctly with a clear problem statement, a reason to act, and a small array of possible solutions. How these options are communicated may well decide which project or policy is implemented. A geographer has valuable expertise and perspectives for communicating such options to high-ranking officials, and thus stands to gain their professional trust and confidence.

OPPORTUNITIES FOR GEOGRAPHERS IN STATE AND LOCAL GOVERNMENT

Geographers are involved in a variety of state and local government organizations and departments that focus on, but are not limited to: transportation, health, environmental planning, monitoring and remediation, metropolitan or rural planning, economic development, and GIS. However, state and local governments do not have a common organizational structure. Some local governments may have certain functions that others lack, and functions found within one city may not exist elsewhere, or may be subsumed by county, regional, or state agencies. Therefore, in hunting for jobs, it is important to look into the way local and state government agencies are organized in the places of interest to you. Many agencies and government offices have websites that outline the services they provide along with relevant contact information. In the following paragraphs we sketch some of the most common areas where geographers find employment in the public sector.

1. ***Transportation agencies and authorities*** plan and provide transportation services for the public. It is up to the local legislature to determine what type of service meets their needs. In some areas, a combination of public and private services is utilized. Typically, these organizations will use maps created from demographic data to determine where to place transit routes, or how to alleviate congestion where the vehicle counts are high. They may also utilize base maps or census demographic maps provided by other government agencies. Many agencies will have their own schedulers and planners, but may rely on government geographers for information related to higher level planning. They vary in sophistication from smaller rural systems to larger countywide or regional systems. Many of the smaller systems will seek technical assistance from geographers at the county or state level.
2. ***Health agencies*** include public or publicly subsidized private organizations that seek to promote the general health and wellness of a service area. This may be a county-based agency or a health network charged with delivery of these services. Often, geographers can map localization or concentration of health issues, such as asthma hospitalization rates, "cancer clusters," or other environmental health data in a geographic area. Geographers that work in the public health area of state and local government often have a background in human geography. They use their skills to understand human activities and interrelationships with the physical environment. Furthermore, geographers practicing in public heath often use quantitative methods to conduct research.
3. ***Environmental agencies*** are charged with the protection and conservation of natural resources and, in some cases, cultural resources. Since the late 1950s, managing environmental quality has become a more prevalent function across all levels of government. The passing of legislation such as the Water Pollution Act (1948), the Clean Air and Water Resources Planning Acts (early 1960s), and the establishment of the EPA (1970) have contributed to the growth of environmental regulations (EPA 2011). Many states, in an effort to make natural resource protection more efficient by consolidating environmental programs, formed state environmental agencies that employ geographers.
4. ***Metropolitan or rural planning departments*** are charged with making short- and long-range urban planning and transportation decisions, based on government and private input. Projects can be as small as creating a new intersection at a major thoroughfare, or as large as building a new airport. They are funded by a combination of federal and state planning money. In rural areas, they are used to plan how transportation might be related to commercial interests such as farming. In larger urban areas, planning departments

are involved in major investments in ports, airports, buses, roads, and railways. Urban planning is a function that many of these planning departments provide, with the goal of improving communities by creating environments that integrate where people live, work, and seek recreational opportunities.

Urban planning helps communities envision their future and find a balance between new development, environmental conservation, and innovative change (APA 2011). Geographers contribute to this mission by engaging in environmental planning, land use planning, and demographic analysis. Often, planning departments rely on standing and ad hoc committees of which a geographer may be a member. Larger metropolitan areas will often have GIS staff in their planning department. Smaller areas may or may not, and some come to rely on a person with GIS skills from another department to help them with complex analyses.

Geographers involved in land use planning and demographic analysis are typically focused on developing models and working with GIS. This information may be integrated into transportation and environmental planning functions to better understand the current environment and where changes may occur in the future. For instance, an understanding of the high-priority environmental resources in the region, along with knowledge of current development and population growth, can assist planners and policy makers in making decisions about how best to protect environmental resources.

5. ***Economic development agencies*** seek to attract new firms and promote existing businesses. Often, they will employ GIS to understand how local economic development issues, laws, and policies affect businesses. In many cases, there are restrictions on building in historic preservation districts, on wetlands, or similarly regulated areas. An economic development agency may ask for maps or analyses of demographics, access to highways, or the availability of transit. As a civil servant and geographer, you may be tasked with creating any of these maps, and as with environmental issues, it is vital to be impartial. The public is sensitive about job growth, especially in struggling areas facing high unemployment or the loss of traditional anchor industries. As such, maps and reports for these agencies must present data carefully so as not to favor any one particular area for development. If the agency favors one location, it's important that this conclusion is drawn from a variety of public inputs. There may also be incentives to place development in certain areas that will likely be politically sensitive.
6. Finally, some state and local governments have established ***departments focused on GIS*** as a means of developing data and mapping systems for use by public agencies or the public in general. Over the past several years, GIS has evolved from specialized software that only a few skilled professionals could use to a tool that a variety of professionals can use to perform analysis in their area of expertise. However, for developing geographic data, models, and applications using GIS, more specialized professionals are still needed. As one illustration, see the profile of Dan Haug, who coordinates information technology (IT) and GIS services for the Confederated Tribes of the Umatilla Reservation in Oregon.

Although the basic capabilities of GIS software may be used across a variety of departments, centralized teams with GIS expertise are now common in many government work environments. These teams focus on tasks such as developing large-scale geographic data, analytical models, and interactive GIS applications that simplify functions so that they are more accessible to non-GIS professionals. They may support the data, modeling, and applications needs of multiple programs within a department or an entire agency. As previously mentioned, there

PROFILE 6.1

Dan Haug, IT Director, Confederated Tribes of the Umatilla Indian Reservation (Pendleton, Oregon)

Dan Haug discovered geography at the age of 19 while working at an Arby's restaurant. Although he had never taken any geography classes, a friend who recognized his strong math and computer skills recommended him for a cadastral mapping project. Dan soon realized he had a knack for spatial thinking and decided to become a geographer.

Dan received his bachelor's degree in geography and computer science from Evergreen State College in Olympia, Washington. Because Evergreen offers a non-traditional education, he was able to complete his junior and senior years via independent study while based in Belize, where his wife was conducting field research for her Ph.D. in anthropology. He did his first formal geography coursework in the master's program at Penn State. Dan observes that a traditional broad geography education provides the ability to **identify the various tools and theories** one can bring into play in order to solve problems. This knowledge is incredibly useful in his job as IT Director for the Confederated Tribes of the Umatilla Reservation, where his central role is to bridge the needs of the organization and the needs of the IT Department. In addition to his GIS skills, being a geographer gives him the **ability to communicate** with people from other fields to ensure that their needs are being met.

Tribal governments have elements of state government and local government, Dan explains, adding that each tribe is unique because of its history of sovereignty. Umatilla's treaty permits traditional uses on portions of its former homeland that are public lands within the states of Oregon, Washington, Idaho, and Montana. While this situation is advantageous to the tribe, it **requires the coordination** of natural resource management, data and mapping, and governance functions across multiple state jurisdictions. Fortunately, the tribe has the staff and resources to not only handle these tasks, but to offer assistance to neighboring governments. For example, the tribe has its own GIS Department with six staff and more than 50 nonspecialist GIS users in other departments, such as fisheries, economic development, and planning.

Much of Dan's job satisfaction stems from the feeling that "what I do on a daily basis has an impact on the lives of the people I work with." But he cautions prospective employees, especially those who are not tribal members, that an overly idealistic or paternalistic mindset is a big mistake. He is quick to note that the job is not "development work" and the tribes don't need outsiders to "bring enlightenment." Increased self-determination means that they often have better resources than other local governments, and Dan contends that the Umatilla Reservation has the most sophisticated government in all of eastern Oregon.

When hiring new employees, Dan believes that **work experience** is just as important as an applicant's academic background, and sometimes even more so. Today he looks for candidates with the right attitude and education as well as hands-on experience. Unsurprisingly, given his background, he firmly believes that the best learning happens outside the classroom. "If you have the opportunity to work on an applied project in school, that's incredibly valuable," Dan observes. "Find something that interests you and spend some of your own time on it to develop marketable skills."

—JOY ADAMS

is typically not a common structure for how these teams work within their respective organizations. Some may attempt to integrate GIS professionals into each program, whereas others choose to have a more centralized approach.

STRATEGIES FOR GETTING A JOB IN THE PUBLIC SECTOR

When an organization hires a person, it typically is looking for a solution to a problem for which it lacks adequate staff. While few companies (aside from land use firms, GIS companies, and similar businesses) will put out a "help wanted ad" explicitly asking for a "geographer," many government agencies will ask for skills that geographers have. So, first and foremost, ask yourself a basic question: "Even if the agency in question isn't explicitly asking for a geographer, are they asking for what I can do?" This, in turn, creates a natural follow up question: "Can I prove that I am the person this agency wants to solve the problems they have?"

All hiring is done for a purpose. In government, there are specified lists of qualified individuals who have passed certain tests for specialized training, certification, and licensure. It is very important for anyone looking into government work to read the exam qualifications and understand them. These qualifications will not only help you understand what training you need to succeed; they will also alert you to the skills government agencies are seeking. After the examinations, there are typically canvass letters that define a specific position and ask for applications from interested and qualified candidates. The letters will give you a further idea of that agency's needs. For instance, a position for a research specialist at a department of health will be far different from a research position at a department of labor. Each agency will have different needs, and in most cases, different required skill sets and backgrounds. So, a geographer looking to start a career in government must do the same kind of fact-checking as a person seeking a position in the private sector might need to do. Again, ask yourself: Do I have the appropriate skills even if the specific canvass letter doesn't ask for a geographer?

Positions in state and local government are usually filled through either internal or external hires. In filling most vacancies, preference is given to those already in government who are qualified for a certain position. If that list is exhausted, and approval for external hiring has been granted, "open" or "competitive" positions may be canvassed. Such positions would be open to all of the current nongovernmental employees who have passed the appropriate civil service test.

One of the authors, in his first attempt to pass a civil service exam for "economist," submitted his qualifications for the exam, including a transcript from his alma mater and a résumé. However, because many of his classes were in geography rather than economics, the initial application to take the exam was denied. Thus, he had to explain each and every economic geography course he had taken. After this explanatory letter was reviewed, he was admitted to the examination, passed it, and subsequently responded to a canvass letter. But even while interviewing for an opening, he sensed that some of his interviewers had trepidations: Did he, as a geographer, really have the knowledge of statistics and economics that they needed? These are the kinds of questions geographers must be prepared to address in detail, especially since few interviewers will be familiar with the field of geography. So, unless you are interviewing for a GIS position, assume that the interviewer will never have interviewed a geographer, that you won't know what to expect, and, accordingly, that you will need to clearly articulate how your expertise fits the position.

If you've done your homework, this shouldn't be too difficult, although you need to come prepared. It is important to know how the position is described in the canvass letter. Then, put together a dossier of your related accomplishments, both in college and in the working world. If you are a new graduate, document your internships and be prepared to show the results of those projects, such as presentations, reports, or publications. If you are still in college or graduate school, do yourself a favor and get a professional internship, even an unpaid one (for more on internships, see Chapter 4 by Denise Blanchard, Mark Carter, Rob Kent, and Chris Badurek).

In government, the purpose of the interview isn't necessarily to assess skills; the canvass and civil service exam have already done that. Rather, the interview seeks to assess fit, commonly called Best Fit or Qualifications (BFOQ). Government hiring tends to focus more on the candidate's overall fitness for a needed role than on "who scores highest" on civil service exams. One of the analogies we've heard is that people don't buy drill bits because they need drill bits; they buy drill bits because they need a hole. Think in the same way when interviewing. A government agency doesn't need you because you have a great résumé, are highly intelligent, or have a degree from a prestigious university. A government agency needs you to solve a problem, and just as importantly, to solve it wisely and well with the people's tax dollars. If you are that person and that solution, then you will be the employee of choice.

EDUCATION AND TRAINING SUGGESTIONS FOR PUBLIC SECTOR GEOGRAPHERS

Most geography positions in government require at least a bachelor's degree (BA or BS) with a concentration in a subfield such as GIS, environmental, or physical geography. But cross-training in other fields is useful as well. You might consider pursuing a minor or some coursework in geology, computer science, ecology, economics, or engineering. At the same time, competition for some jobs has increased to the point that a master's degree may be required to secure or advance in a position. Although the Bureau of Labor Statistics (BLS 2010b) forecasts significant growth for "geographers," "geoscientists," and "computer and information scientists," growth in these fields will include many positions requiring higher levels of skills.

We encourage those with a passion and interest in geography to continue into graduate school. In addition, a professional internship demonstrates to a potential employer that a person has experience in a workplace environment. Both a master's degree and an internship offer major advantages in a tight labor market. The cost of training new workers can amount to $3,500-$5,000 or more per year (Sasha Corporation 2011), mostly for the technical training needed on the job. If informal training were included—the type that occurs on the job and often develops knowledge and insight as to how a professional environment works—the cost would be even higher. A person who has completed a professional internship has a head start on both types of training (Frazis and Loewenstein 2003). Both of us have seen interns hired into full-time positions immediately after graduation or even while finishing their degrees.

After you achieve your first geography position, continue developing your skills. Take advantage of any training the employer will pay for, and also consider pursuing the free online training opportunities offered by many vendors. It is absolutely critical in today's government sector, much as in the private sector, that new skills be developed constantly. In government, a person who fails to develop skills relevant to new realities, especially in a technical field like GIS or geographic analysis, runs the risk of falling into a career rut or being sidelined in advancement. Moving upward in government is more a function of your ability to develop new knowledge and skills than it is of what you learn in college and graduate school. This point is underscored by Moncarz (2002, 45): "Technology changes at such a rapid pace that retraining and updating information technology skills is essential, even for workers already in their jobs."

"SOFT SKILLS"

So-called soft skills are the general abilities important to professional work, which include developing interpersonal relationships, leadership, coaching, team-building skills, friendliness, and optimism. Much of our education and training focuses more on "hard skills"—that is, the knowledge and abilities we develop as geography majors or graduate students. The following is a short discussion of the general skill areas you should develop as preparation for a public sector career. For additional context, be sure to read Dmitry Messen's profile for a comprehensive look at the value of geographic and general skills for research-oriented work in the public sector.

1. ***Written and Oral Communication.*** Your skills as a geographer are only as useful as your ability to communicate effectively. As Tittel (2008, 2) has written in regard to information technology (IT) careers,

   > It's no exaggeration to say that written communications are important to nearly every job. This is especially true for IT, where reports, analyses, evaluations, and even plain, humble e-mail messages, are part of the regular grind. Though it's rare to hear of IT professionals garnering promotions or new jobs solely on the basis of writing ability, a knack for clear, cogent communication never hurt anybody's prospects.

   Throughout your career you will share your work through oral presentations, written reports, maps, memos, and briefing papers. In many instances the audience with whom you will be communicating will not be experts on the topic you are presenting. It is important, then, to learn as much as possible about your audience in advance so that you can organize presentations and reports suited to the level of audience interest and expertise. You may also have to communicate controversial topics, highly political issues, or projects with a lot of money on the line—and the people you address may be hostile or even angry about the issues discussed. This is all the more reason to make sure that you are prepared with the facts and have carefully prepared your report, tables, and maps. It is often in such circumstances that a geographer's extensive knowledge of the situation of a local project can come in handy.

   Effective writing isn't a talent that develops overnight; it improves with practice and experience. The same holds true for oral communication and presentation skills. For many, conducting presentations is an unnerving experience no matter the size of the audience. Others relish such opportunities. In either case, we encourage you to gain as much experience and practice as possible in making oral presentations. Although the senior members of most organizations are usually called on to give more presentations than new hires, the ability to speak well in public should be cultivated from the very start of your career. If oral communication is not one of your strengths, there are books and classes that may be useful, as well as groups such as Toastmasters International that hold local workshops.
2. ***Problem Solving and Analytical Thinking.*** Many geographers who are employed at state and local levels are involved in efforts that analyze data or work to provide solutions to problems of the day. A professional's ability to solve such problems and to analyze information effectively will allow him or her to stand out as a high performer. Problem solving and analytical thinking, coupled with a person's ability to communicate findings, form a valuable combination. However, be aware that solving problems involves not only applying one's own knowledge, but also knowing when to seek the help and advice of others.
3. ***Project Management.*** Project management is a multifaceted skill set and is perhaps as much an art as a science. The science behind project management is a person's ability to develop project plans, manage day-to-day activities, and achieve project goals. But the art

PROFILE 6.2

Dmitry Messen, Program Manager, Socio-Economic Modeling Group, Houston-Galveston Area Council (Houston, Texas)

Dmitry Messen was first introduced to geography as an undergraduate student at Moscow State University. He had always been interested in the social sciences, despite the "highly political" nature of related employment opportunities during the last days of the Soviet Union. When one of his professors suggested that he major in economic geography, Dmitry was intrigued. "The geography program had an international relations focus," he recalls. "I thought this was a very attractive avenue to an education." After graduating in 1992, Dmitry moved to the United States to pursue a Ph.D. in geography at Louisiana State University. He finished school in 1997 and continued to work as a Research Associate in LSU's Center for Energy Studies, in addition to working part-time in research and consulting positions before landing his current job at the Houston-Galveston Area Council (H-GAC) in 2004.

H-GAC is a voluntary association of local governments within the thirteen-county Gulf Coast Planning Region of Texas. Similar organizations are few and usually limited to large metropolitan regions, so Dmitry's job is rather unique. He is the Program Manager of the Socio-Economic Modeling Group, the "information and research hub" of H-GAC's Community and Environmental Planning Department. "We collect, process, and analyze socioeconomic and land use data for the greater Houston region," Dmitry explains. "Our main product is the long-range, small-area forecast of population, employment, and land use. Additionally, we do a lot of on-demand mini-studies related to **demography, regional economics, and land use** in our region." The data are also made available to other public, private, and government organizations.

Dmitry's position requires a **specialized skill set** including GIS, cartography, statistical analysis of large datasets, computer programming, and database management, as well as knowledge of local geography. "It's the quintessential geographer's job," he observes. "We are working with a spatio-temporal system and the integrated landscape, and we pay no attention to disciplinary boundaries." The group's work straddles the line between the production-oriented focus of government and the research and development orientation of academia. Such positions provide an attractive option for geographers who are looking to transition from traditional academic careers to applied careers.

When evaluating prospective employees, Dmitry considers internships and relevant work experience to be an asset but not essential for those who have successfully undertaken challenging coursework. A solid educational background is the "first threshold" for employment in his group: half of his team members hold Ph.D.s, while the other half hold master's degrees. He advises students to "diversify as much as possible." "Geographers are in the best position to gain knowledge from other fields," he notes, adding that "**cross-pollination**" across disciplines and employment sectors will make candidates more attractive to employers. Perhaps most importantly, job applicants must also have a few intangible qualities, particularly intellectual curiosity and a genuine interest in the work itself. As Dmitry's experience demonstrates, careers beyond the ivory tower can provide great opportunities for geographers to do **cutting-edge applied research**. While positions like his are rare, the upside is that few applicants possess the diverse skills required, so there is not much competition for geographers who develop the necessary qualifications.

—MARK REVELL AND JOY ADAMS

of project management comes with experience: How do you sense when your project is on or off track? How do you effectively facilitate a meeting with people of opposing viewpoints? Is your subject matter a sensitive topic in the community, and if so, how do you convey your results in a manner that is fair and open? How do you motivate your team? These are the realities of managing projects, and the skills involved go far beyond just tracking tasks and writing status reports.

Most organizations require periodic status briefings on the progress of projects and critical tasks. Initially, managing projects that have multiple objectives, deliverables, and milestones can prove daunting. However, there are many software tools on the market that can assist you. Look for those tools that allow you to organize tasks, assign resources and dates, and track progress.

4. ***Ability to Work in Teams.*** Organizations today are realizing that cross-disciplinary teams from different departments are perhaps one of the more efficient and effective ways in which to work. As state and local governments seek ways in which to do more with fewer resources, the idea of temporary project teams is gaining ground. This cross-disciplinary approach is, of course, necessary in many situations involving complex projects that require expertise from a variety of fields. An environmental project may, for instance, require a biologist, hydrologist, environmental scientist, urban planner, and GIS professional to work together. Each member of the team brings complementary skills to the table; no one person can complete the project alone. The ease with which teams leverage each other's skills and apply them to a project is a hallmark of today's work environment. Teamwork also involves communicating effectively and facilitating discussions. If you haven't realized it by now, all of these skills are interrelated. They are tools waiting to be used for the right opportunity.
5. ***Technical Proficiency.*** As we noted earlier, there is no substitute for maintaining the best skills in the industry. Continuous training and development in your government agency's GIS software of choice, in addition to classes related to the specifics of your job, are critical to proving value. Take advantage of every reasonable training opportunity available to you.

CONCLUDING THOUGHTS

By now you know what geographers do, where they work, the skills they apply, and some strategies for finding a job in state and local government. One of the more important things you can do to develop your career as a geographer is to network among colleagues (see Tina Cary's chapter on professional networking, Chapter 5). Networking provides an opportunity to build important relationships, share information, and develop your skills through professional meetings. The AAG's national and regional meetings, the Applied Geography Conferences, those of local or regional GIS associations and other professional memberships, and your collegiate alumni association all provide important potential contacts. If you keep these up to date and active, you may have an opportunity to exchange information, find new resources, and improve your reputation as a geographer in civil service.

Finally, we encourage you to explore the area of government in which you are most interested in developing a career. Get to know professionals within these areas of government, and try to be at the top of their list of potential candidates when new opportunities arise. A proactive approach to managing your career can go a long way toward developing unique career opportunities.

REFERENCES

APA (American Planning Association). 2011. What is planning? http://www.planning.org/aboutplanning/whatisplanning.htm (accessed 24 June 2011).

BLS (Bureau of Labor Statistics), U.S. Department of Labor. 2009. An occupational analysis of industries with the most job losses. http://www.bls.gov/oes/highlight_jobloss.pdf (accessed 24 June 2011).

BLS (Bureau of Labor Statistics), U.S. Department of Labor. 2010a. State and local government, except education and health. http://www.bls.gov/oco/cg/cgs042.htm (accessed 24 June 2011).

BLS (Bureau of Labor Statistics), U.S. Department of Labor. 2010b. Employment projections. http://www.bls.gov/emp (accessed on 24 June 2011).

Crosby, O. 2005. Geography jobs. *Occupational Outlook Quarterly* 49 (1): 2–17. http://stats.bls.gov/opub/ooq/2005/spring/art01.htm (accessed 24 June 2011).

EPA (Environmental Protection Agency). 2011. About EPA. http://www.epa.gov (accessed 21 October 2011).

Frazis, H., and M. A. Loewenstein. 2003. Reexamining the returns to training: Functional form, magnitude, and interpretation. Washington, DC: Bureau of Labor Statistics, U.S. Department of Labor. Working Paper 367. http://www.bls.gov/osmr/pdf/ec030040.pdf (accessed 24 June 2011).

Moncarz, R. 2002. Training for techies: Career preparation in information technology. *Occupational Outlook Quarterly* 46 (3): 28–45. http://stats.bls.gov/opub/ooq/2002/fall/art04.pdf (accessed 24 June 2011).

Sasha Corporation. 2011. Compilation of turnover cost studies. http://www.sashacorp.com/turnframe.html (accessed 24 June 2011).

Tittel, E. 2008. The importance of soft skills. http://itknowledgeexchange.techtarget.com/it-jobs/soft-skills-part-2-of-4-written-communications/ (accessed 24 June 2011).

7

Emerging and Expanding Career Opportunities in the Federal Government

Allison M. Williams, Molly E. Brown, Erin Moriarty, and John A. Wertman

Geography is an interdisciplinary field that provides a solid foundation for government careers at the national level. Geographers are employed as scientists, researchers, administrators, resource planners, policy analysts, project managers, and technical specialists across a wide variety of federal agencies. In fact, geographers are employed in such a great range of positions at the national level that we can only sketch out some of the major opportunities in this chapter. All of these opportunities relate to geography's value across a wide array of government careers.

Geography provides a conceptual framework for analyzing and understanding complex relationships between people and places over time and space, together with tools for measuring, predicting, and reporting change and outcomes. With the call for more evidence-based policy, accountability, and transparency in the public sector, geography is playing an increasingly central role in informing efforts by agencies at all levels of the federal government. Written from both American and Canadian perspectives, this chapter will introduce you to government careers generally before discussing specific opportunities in a variety of federal agencies and what you might do to prepare for a career at the national level.

CONSIDERING A CAREER IN THE FEDERAL GOVERNMENT

In both the United States and Canada, the federal government has numerous, varied, and a constantly growing number of geography-related positions due to the expanding need for geographic expertise and analysis. So what attracts geographers to these types of careers? First and foremost, careers in the public sector offer geographers an opportunity to make a direct impact at national, state or provincial, or local levels, a point also raised by Bass and Quodomine in Chapter 6 on career opportunities in state and local government agencies. A desire to contribute to our city, state, or nation—to "give back" to our community or country—is an important motive for a career in the public sector.

Equally important for many geographers are the unique intellectual, scientific, and personal opportunities presented by working at the federal level. Geographers are involved in developing management plans for national parks, forests, wilderness areas, wetlands, and waterways. For example, they work in the National Aeronautical and Space Agency (NASA) and the National Oceanic and Atmospheric Administration (NOAA) designing, launching and using scientific data from world-class satellite remote sensing instruments to examine the weather and climate variability around the world. They provide robust evidence for policy making at the U.S. Bureau of the Census and Statistics Canada. Geographers are also planning for and mitigating the impact of disasters in the field of emergency management. These jobs involve work that is challenging and complex which present opportunities at scales that might never be available in the private sector, or while working at the state or local level.

Government agencies are an excellent place to work, offering an enormous diversity of positions and opportunities. The public service in Canada is not only a national employer, with over 1,600 points of service across the country, but it is also a wide-ranging international employer, presenting opportunities in more than 150 countries with over 1,000 different types of jobs (Public Service Commission of Canada 2008). The U.S. federal government alone employs more than 1.6 million full-time, permanent civilian workers, and it is worth noting that federal jobs are available in all fifty states and around the world (Bureau of Labor Statistics 2011). Approximately 85 percent of all federal employees work outside of the Washington, DC area. The ability to work in a diversity of places and the freedom to move as new career opportunities arise is an appealing aspect of government work.

Even in this era of economic uncertainty, there are likely to be a growing number of government job opportunities for geographers. This results from the aging of the federal workforce, a trend that is forcing agencies to plan for the replacement of large numbers of retirees in the years ahead. The U.S. government projects that 241,428 federal employees will retire by the end of fiscal year 2012 (30 September 2012), and one out of every three federal employees is expected to retire by 2018. Even in an era of declining government budgets, these statistics will lead to outstanding opportunities for geographers in the coming years (Bureau of Labor Statistics 2011).

You may wonder what it is like to work in a federal government agency. Federal employment has a number of advantages. Positions generally offer a high level of job security and are often much more stable than those in the private sector. With hard work and commitment, you can advance your career in a single agency for much longer than is common today in the private sector. In addition, once you are in a federal position, it is much easier to move both within and between agencies, often without an extensive search process. This means that once employed, you will have access and exposure to opportunities in any department, from education to international relations.

Federal pay schedules can be very competitive with the private sector. Conditions of work and benefits such as retirement and health plans, vacation time, and opportunities to arrange flexible daily or weekly schedules also support life beyond work in ways that are not always offered by private sector employers. Jan Monk reinforces this important point in her discussion of work-life balance in Chapter 14.

Federal government jobs also offer opportunities for advancement comparable to work in the private sector. In the Canadian federal government, for example, there are over thirty-six CEO equivalent roles and 4,500 executive positions. To be a competitive employer, the Canadian government seeks to develop capacity and retain top employees through programs offering a range of professional and personal development opportunities and support, such as those noted in Chapter 16 where Kneale and Maxey discuss lifelong professional development.

You may also be wondering how workers in federal agencies are evaluated for promotion. This system varies by agency, but to offer one U.S. example, some scientists at NASA are required to demonstrate independent scientific scholarship as a key criterion for promotion. NASA scientists are also evaluated based on the impact of their scientific research, their accomplishments and responsibilities, their leadership, the value of their professional service to the agency and the scientific community, customer satisfaction, relevant experience, and many other factors (e.g., educational attainment, time-in-grade or time-in-position).

Despite the advantages we have identified, some aspects of working for the federal government may not appeal to you. These include a bureaucratic work environment and a high level of management oversight, which may mean some processes take longer than expected. Leadership of agencies by administration appointees can also lead to the politicization of certain issues and policy decisions, which you may disagree with professionally or personally. Navigating the large number of federal rules and regulations and learning the "unwritten" culture of many federal positions, including procedures and information flow, can be daunting for newcomers. Understanding how work gets done in federal work environments can take some time, but with good advice and persistence, a challenging and robust career is certainly attainable. Another thing to keep in mind if you are looking for opportunities in the Canadian federal government is that for management positions, you will likely need to be fluent in French and English. Depending on the region where you work and the sector within which you work, you may need to be competently bilingual.

THE GOVERNMENT PERSPECTIVE: WHY HIRE GEOGRAPHERS?

Now that we have discussed some of the characteristics of a government career, you may still be wondering, "But why would they want *me*?" With the recognition of the growing scale and complexity of social, environmental, political, and economic challenges, it is becoming apparent that addressing these issues through evidence-based policy making will require rigorous research, analysis, and evaluation. Geographers and geospatial experts have the skills needed to carry out this critical set of tasks. Furthermore, with the growing need for evidence, collaboration, interdisciplinary research, and knowledge transfer, there is a clear role for geography and geographers across many agencies (James et al. 2004). As Banks (2009, 15) has stated: "You can't have good evidence, you can't have good research, without good *people*. Whether it is in the public, private, or not-for-profit domain, there is a need for qualified and skilled researchers who are able to use their technical expertise as well as a holistic approach for understanding and resolving complex issues."

Broad-thinking geographers have this holistic approach and are uniquely qualified to step into a variety of positions. Within the Canadian context, for example, graduates have secured careers as policy analysts with numerous different federal departments, including Environment Canada and Canadian Heritage. Their work takes many forms. The Canadian Environmental Assessment Agency, for example, provides policy analysis and advice to protect the environment. On the ground, this agency has contributed to innovative solutions to protect sensitive ecosystems and put forward sustainability measures for major development projects, such as the expansion of the Whistler Nordic Centre in British Columbia for the 2010 Winter Olympic and Paralympic Games (Canadian Environmental Assessment Agency 2011). Many agencies and departments within the Canadian federal government target the improvement of the Canadian people's economic and social well-being. This objective is achieved by removing barriers to participation, building capacity through training and education, and developing fair and equitable policies to support all members of society. Social geographers are applying their skills in numerous areas, such as in conducting gender analysis with the Status of Women Canada, a federal government organization dedicated to promoting the full participation of women in society.

Increasing globalization has impacted the interdependency of localities, as witnessed in the global economic crisis, pandemic events, and severe weather occurrences. With these challenges come exciting opportunities to develop and implement innovative solutions. This is evident in the growing field of emergency management, where the Canadian federal government provides leadership and guidance to all levels of government. Public Safety Canada, the Canadian national body for emergency management, develops national policy and builds capacity to prepare, respond, and recover from critical incidents. Geographers are contributing to the emergency management field in numerous ways, such as by mapping hazards and developing notification systems, as well as contributing to disaster mitigation, emergency management planning, and training (Public Safety Canada 2011). Furthermore, physical geographers have successful careers in informing this work with Environment Canada, whether working on oil spill detection or on sea ice and lake ice analysis.

GEOGRAPHY IN FEDERAL GOVERNMENT: A LONG AND VARIED RECORD OF CONTRIBUTIONS

Geographers are employed widely within federal government agencies, not only in the United States and Canada but in other countries as well. In many cases, this participation dates back to some of the first national surveying and mapping projects of the 18th and 19th centuries and has expanded steadily since then into a far greater range of careers. Rather than attempt to provide an exhaustive list, this section will highlight some of the agencies where geographers have been and often continue to be employed most extensively.

The United States Geological Survey (USGS), founded in 1876, was perhaps the first true home for professional geographers in the federal government. Today, the USGS, which is part of the Department of the Interior, employs a significant number of professional geographers—and research conducted by individuals such as Dave Selkowitz, whose work is profiled here, lays the groundwork for significant policy and regulatory actions. For example, USGS geography research on the earth's climate and the related topics of global change and species migrations has contributed to decisions related to U.S. climate change policy. In addition, USGS geographers have played a key role in the ongoing series of ACES (A Community on Ecosystem Services) conferences aimed at "advancing the use of ecosystem services and related science in conservation, restoration, resource management, and development decisions." The Association of American Geographers (AAG) has also been a partner in ACES, which brings together leaders from the government, private sector, academe, and NGOs (nongovernmental organizations) to discuss and lay the groundwork for action on important ecosystem issues.

Carl Shapiro, Co-Director of the USGS Center for Science and Decisions, asserts that "Geography is a great synthesis science. It plays an important role in land use decisions and resource management by providing a framework for integrated spatial data analysis that incorporates both the natural and social sciences." Shapiro notes that "the Fish and Wildlife Service, the Bureau of Land Management, the National Park Service, and the Bureau of Indian Affairs rely on geographic data in making these refuge, resource, and land management decisions" (Shapiro 2010).

PROFILE 7.1

Dave Selkowitz, Research Geographer, U.S. Geological Survey, Alaska Science Center (Anchorage, Alaska)

Dave Selkowitz has been gradually migrating north and west since receiving his bachelor's degree from Vermont's Middlebury College in 2001. He lived, worked, and studied in Montana and Oregon before joining the staff of the U.S. Geological Survey's Alaska Science Center. "Alaska is an amazing place," he says. "There's nowhere else like it."

Dave's job involves conducting **research on land cover, vegetation structure, and snow cover**. He is usually responsible for multiple projects simultaneously. When working in the office, he spends most of his time processing remotely sensed imagery, analyzing field data, and writing manuscripts for publication. In the field—where he spends anywhere from three weeks to three months per year—he collects vegetation and snow cover data from sites throughout the western United States and Alaska. His work involves a great deal of travel, because his primary study site is the entire state of Alaska. He presents his research at several conferences each year.

Dave recalls that his interest in geography began "when I realized that geographers were the type of scientists who study the things I was really interested in." Growing up, he was an avid skier, and he found himself fascinated by how patterns of snow and vegetation vary with topography. He wrote his undergraduate thesis on avalanches and climate in the western United States. He then worked for a U.S. Geological Survey research group in Montana before entering the graduate program at Oregon State University. After completing his master's degree in geography, Dave's outdoor interests drew him to Alaska, where he soon landed a job **working as a contractor** for the USGS and eventually secured a full-time, permanent position with the agency. "Having USGS contacts and understanding the expectations of the job definitely made for an easier transition," he explains.

Working in **rugged and often remote environments** often requires skills beyond those developed through traditional academic training. For example, Dave points out that he didn't learn how to avoid bear attacks in graduate school, but fortunately, the USGS provides backcountry safety and firearms training programs for all fieldworkers. In addition, his previous recreational activities, such as skiing, hiking, rafting, and outdoor navigation, all come in handy while collecting field data. Having grown up in Connecticut, Dave says he is not bothered by the cold, but the lack of daylight can be very hard. His frequent business trips are therefore especially welcome during the long winter months.

In the fall of 2011, Dave will begin the Ph.D. program in geography at the University of Utah. As a participant in the federal government's **Student Career Experience Program (SCEP)**, he will work at least half-time with the USGS while completing his coursework. The program enhances participating students' learning through relevant work experience while providing federal employers with an opportunity to take an active role in developing their future workforce. (For more information on SCEP, see http://www.usajobs.gov/ei/studentcareerexperience.asp.) Dave will work full-time during the summers, and he plans to return to Alaska after receiving his degree.

Although climate change might be bad news for the planet, there is a silver lining for geographers like Dave because agencies such as the USGS will have an important role to play in monitoring and understanding its impacts. "We need to understand these things and why they are changing," he observes. "So, geography is likely to be in high demand through the foreseeable future."

—MARK REVELL

For many years, geographers at the USGS worked together in the agency's Geography Division. In August 2010, however, USGS undertook a major reorganization, and the key geography program activities were moved into new branches on Climate and Land Use Change, Core Science Systems, Energy and Minerals, and Environmental Health. The realignment is based on the agency's 2007 science strategy document and should provide a boost to geography research given the field's interdisciplinary nature.

Although geographers continue to play an important role at the USGS, there are many other U.S. federal agencies, laboratories, and research centers that employ geographers, such as the Oak Ridge National Laboratory, an applied science and technology research center in Tennessee; the National Renewable Energy Laboratory in Colorado; the National Park Service; and the U.S. Forest Service. The U.S. Census Bureau is among the most important of these agencies. Scientists and professionals in the Geography Division conduct research and related projects that play an important public policy role. The Geography Division, for example, works with state and local governments to identify boundary changes and groups Census data into units known as Census Tracts, Block Groups, Blocks, and Census-Designated Places. These groupings are a key part of federal, state, and local efforts to divide and allocate billions of dollars in government funding for a huge number of programs. Population data developed by the Geography Division is also extremely important to state efforts to redistrict their congressional and state-level legislative districts. For additional insights into the contributions of geographers working at the Census Bureau, see Mike Ratcliffe's profile.

Similar policy-related research is done at Statistics Canada, where quantitatively trained human geographers are employed, using their skills in the creation and analysis of census data for informing public policy. Statistics Canada has a group that specializes in geography products, where information about the location, shape of, and relationship among geographic features, such as road networks and boundary files, is analyzed and made available.

Another major federal responsibility that has historically relied on the work of geographers is in the area of foreign policy. The State Department opened its Office of the Geographer in 1921, and today the office includes approximately 35 full-time staff, some of whom are detailed to the department from other federal agencies. The office also relies on the work of visiting fellows who come from academic settings. While the office's chief function is to ensure that all official government maps reflect the foreign policy standards and agreements of the federal government, in practice the staff does much more than that. For example, providing information on the social and humanitarian context that complements the maps and images distributed by the office is a critical role that can help shape foreign policy considerations—especially for unsettled world regions (Taylor 2009).

The office is organized according to the needs of the State Department's various policymaking bureaus. According to Taylor (2009), however, "in the mid-1980s, the office expanded to address problems that did not neatly fit the diplomatic needs of the Department's regional bureaus—matters such as refugees, human rights, democratization, and international environmental concerns." The work of the geographers in the office often requires them to be deployed overseas—sometimes related to the diplomatic priorities of the State Department, but also frequently for joint missions with other federal agencies, or even international organizations such as the United Nations. Canadian geographers are involved in similar work across numerous departments, such as the Canadian International Development Agency, the Immigration and Refugee Board of Canada, Foreign Affairs and International Trade Canada, and the Foreign Service.

The work of State Department geographers ties in closely with that of their counterparts in the National Geospatial-Intelligence Agency (NGA), an arm of the Department of Defense that was originally created as the National Imagery and Mapping Agency in 1996. While the NGA's

PROFILE 7.2

Mike Ratcliffe, Assistant Division Chief, *U.S. Census Bureau* (Washington, DC)

Mike Ratcliffe likens growing up in the Washington, DC area to growing up in a company town: "Many of the adults you encounter work for the government, so working for the government becomes the thing you do after college." His father and grandfather both worked for the Department of Agriculture, and his uncle was a biologist for the Food and Drug Administration. As a political science major at George Mason University, Mike had originally envisioned a career in public policy or law. But after a geography professor told him, "If you're not already a geography major, you should become one," he began developing an interest in the **relationship between geography and data** and how data could be used in policy making.

Mike's career began at National Geographic, where he landed a clerical job in the illustrations library shortly after completing his bachelor's degree. After completing his master's in geography at Oxford University, he intended to return to National Geographic, but a hiring freeze left him searching for a new job. After sending his résumé and cover letter to every agency listed in the AAG's geography contact list for the DC area, he was soon contacted for an interview by the U.S. Census Bureau's Geography Division Chief. While he admits that at the time he had "little familiarity with the Census Bureau," he was offered a job.

Since joining the Census Bureau in 1990, Mike has **worked his way up through the ranks** to his current position as Assistant Division Chief for Geocartographic Products and Criteria. These products include TIGER files, thematic and reference maps, and geographic reference information available on the Census Bureau's website. In previous positions in the Geography Division and Population Division, Mike developed definitions and criteria for urban areas, metropolitan and micropolitan statistical areas, and other statistical geographic areas to aid in the tabulation and dissemination of data.

Mike notes that geographers are quite common among Census Bureau staff, so many of his co-workers have some formal training in, or at least a basic familiarity with, GIS software. However, technical information must be made accessible to users, so **excellent communication skills**, and especially writing abilities, are very important to his work: "Census Bureau employees need to be able to explain concepts to nongeographers. The ability to take a geographic concept and present it to the lay public in a way they can easily understand is absolutely critical." As much of the bureau's work requires a thorough knowledge of U.S. geography, he adds that "we look for people with **good, traditional geography skills** as well."

While Mike notes that employment prospects in all areas of the federal government depend on congressional funding priorities, he believes the long-term outlook is positive due to a large number of impending retirements. Furthermore, he believes that opportunities for geographers will be sustained by the need to continually update and maintain geospatial data: "Geography will always have a central role in everything the Census Bureau does."

—MARK REVELL

budget and many of its functions are classified, the agency works closely with State, Defense, and the U.S. intelligence community to ensure that the government has the surveillance and imagery expertise needed for critical foreign policy and military decision making. The agency relies heavily on geographers, especially those with expertise in geospatial analysis and geographic information systems (GIS), as well as foreign-language knowledge and foreign-area expertise. Current NGA Director Letitia Long has also expressed a desire to augment the agency's human geography expertise, noting: "The robust incorporation of human geography into our analytic tradecraft combined with the use of technology will help us make sense of larger and larger volumes of data" (Long 2011). Long also ties the work of human geographers into key agency planning and resource allocation decisions. Government work for geographers can involve the most sensitive and critical aspects of national security and prove an exciting challenge.

While geographic information systems have become an irreplaceable tool for use by the intelligence community, GIS is also used extensively by almost every federal agency in the American and Canadian governments. The Department of Transportation and Canada's equivalent, Transport Canada, for example, rely on geographers and GIS experts as they undertake analyses of road projects, highway safety, or heavily traveled aviation routes. NOAA—which employs a large number of geographers working on climate and other key issues—uses GIS to solve ecosystem management questions.

Fish habitats, including habitat restoration issues, the tracking of marine mammals, and the analysis of commercial fishing patterns, are all covered at NOAA using GIS for spatial analysis and understanding of ecosystem functioning. Jamie Brown Kruse (2010), NOAA's Chief Economist, notes that GIS "provides a critical background for national ocean policy decisions and the federal response to hazards issues." She also explains that the Obama administration has directed federal agencies to maintain high standards for scientific integrity and that scientific justifications for policy decisions are a critical asset in government transparency efforts. The agency's use of GIS and its reliance on the expertise provided by its professional geographers on staff play a key role in meeting the administration's goals.

One vital aspect of federal policy making that is only beginning to take advantage of the full power of GIS analysis is healthcare policy. As the Obama administration's federal healthcare reform law is implemented and electronic medical records systems become commonplace throughout the nation, the Department of Health and Human Services (HHS) will become increasingly reliant on GIS and geographic analysis. Already, several key HHS agencies, including the Centers for Disease Control and Prevention (CDC) and various components of the National Institutes of Health (NIH), employ geographers who work on population health and related issues. The AAG is working with staff at several NIH institutes to encourage the agency as a whole to take better advantage of the power of GIS as a scientific tool. For example, the NIH could use GIS to study and analyze infectious-disease outbreaks or the difference in health outcomes for rural and urban communities facing similar underlying disease conditions. Within Canada, geographers employed with the Public Health Agency of Canada similarly conduct disease tracking and research on population and public health. Geographers and GIS experts will increasingly be called upon for these critical tasks.

Geographic analysis is also increasingly being used to link economic and environmental considerations. Malka Pattison of the U.S. Department of the Interior's Office of Policy Analysis stated that, "As geography is applied to economic analysis, GIS becomes increasingly useful" (Pattison 2010). She noted that federal analysts and stakeholders rely on GIS analysis to understand the critical dimensions of an issue and the options available to policy makers. Pattison also explained that resource management decisions can be improved by incorporating the results of GIS analysis.

Professional geography in Canada preceded the creation of the Canadian Association of Geography (CAG). During WWII, Hugh L. Keenleyside had served as a first secretary in the Department of External Affairs and secretary to the Permanent Joint Board on Defence, and was a member of the Northwest Territories Council in the 1940s. Convinced of the value of geography to the conduct of politics and international affairs, he saw to it that a geographer was contracted to undertake much-needed northern research. It was in this way that the first professional geographer was hired in the service of the Canadian federal government. When appointed Deputy Minister of the Department of Mines and Resources in 1947, Keenleyside decided that a Geographical Bureau was needed within it. Soon thereafter a Branch was established, the purpose of which was to "collect, organize and make readily available geographical data on Canada and foreign areas of importance to Canada." By 1950, a staff of 10 geographers created the pool from which several of the fathers of the CAG were drawn (Canadian Association of Geographers 2009).

Today, Canada's federal government employs geographers throughout a wide range of departments, most notably in: Environment Canada, Transport Canada, National Defence and the Canadian Forces, Indian and Northern Affairs, National Resources Canada, Fisheries and Oceans Canada, Health Canada, and the Canadian Environmental Assessment Agency. Natural Resources Canada manages the National Atlas of Canada, a public resource that is linked to *GeoConnections* (http://www.geoconnections.org/en/index.html), which is an innovative program that supports projects enabling decision makers to use online location-based information to benefit public health, safety and security, the environment, and aboriginal communities. The 2010 Federal Budget announced $11 million in funding over the next two years to continue to support the *GeoConnections* program and the ongoing development of the Canadian Geospatial Data Infrastructure (CGDI), and provide consolidated geographic-related information to Canadians via the Internet.

A FOCUS ON GEOSPATIAL CAREERS IN THE FEDERAL SECTOR

As already discussed, GIS data and skills are key to the operations of a federal government. The U.S. federal government is a large and growing user of geospatial information. Earlier we mentioned some agencies where use of geospatial data and techniques is growing, but it is important to note that the Federal Geographic Data Committee's 2006 annual report stated that as much as 90 percent of government information has a geospatial information component. Furthermore, the Geospatial Information and Technology Association reports that up to 80 percent of the information managed by the private sector is connected to a specific location. This means that the role of geospatial technologies is likely to grow considerably as this potential is tapped (GITA 2008).

A recent study by the Center for Strategic and International Studies estimates that geospatial-related companies generate at least $30 billion annually (COGO 2010). The geospatial sector has steadily increased by 35 percent a year, with the commercial side growing at a phenomenal 100 percent annually. The U.S. Department of Labor has identified the geospatial sector (both public and private) as one of the three technology areas most critical to job growth. These trends demonstrate why the government will have a strong demand for geographers and geospatial technology specialists in the coming decade (Bureau of Labor Statistics 2011).

The Government of Canada is also moving rapidly to take advantage of the power of geospatial technologies. The Geological Survey of Canada is the premier Canadian agency for geoscientific information and research, with a focus on the highly prioritized areas of sustainable natural resource development, environmental protection, and technological innovation (Natural

Resources Canada 2008). Geographers are contributing to economic development opportunities within the oil and gas sector, as well as the protection of valuable resources such as groundwater. Similar to the U.S., there are expanding areas of exploration for geoscience, such as in the areas of climate change, natural hazard assessments, and public health risk reduction.

Opportunities for geospatial analysis can also be found in other levels of government and in the private sector for the purposes of land use planning, real estate, and emergency management, to name a few. For example, the significant decline of the American eel population in Canada and the United States over the past thirty years has been a focus of research using geospatial data. To determine the cause of the decline and identify actions to be taken, researchers have turned to geomatics and data from the GeoBase National Hydro Network. Fisheries and Oceans Canada, in collaboration with the Quebec Department of Natural Resources and Wildlife and the Ontario Ministry of Natural Resources, has developed a geographic decision-support system (GDSS) for evaluating sites where the American eel no longer has free access to its natural habitat. Further, according to Fisheries and Oceans Canada, the use of geomatics is now critical to allowing a visual analysis of the impact of the various dam construction scenarios.

Geospatial skills taught in geography programs often use quantitative information, with a spatial component derived from a variety of different sources—geology, biology, satellites, ground observations using the Global Positioning System (GPS), traditional surveying, Census data, and others (GITA 2005). Thus, geographic databases make spatial information available in meaningful forms to a wide variety of users, including engineers, scientists, facility managers, resource planners, lawyers, urban planners, and the military. This diversity of uses makes geography a particularly flexible degree for those interested in public service. See the profile of Sarah Brinegar, a Social Science Analyst for the U.S. Department of Justice, for an interesting look at how GIS is assisting the government's efforts to protect voting rights.

In 2010, the U.S. Department of Labor officially defined eight new categories of geospatial jobs (Bureau of Labor Statistics 2011). This comprehensive list provides a firm foundation for describing and tracking these jobs in the federal sector. The Department of Labor also produced a geospatial technology competency model "as a resource for career guidance, curriculum development and evaluation, career pathway development, recruitment and hiring, continuing professional development, certification and assessment development, apprenticeship program development and outreach efforts to promote geospatial technology careers" (DOLETA 2010). Now that the occupational categories and the training model are available, more federal jobs will be defined as fundamentally geospatial, and the path for entry into the federal civil service will be facilitated.

THINKING AND PLANNING AHEAD

So what if you are a student or a recent graduate interested in pursuing careers at the federal level? Look for internship and summer opportunities to get your foot in the door and begin accruing federal government experience. These opportunities may include, for example, job shadowing, mentorship programs, and a wide range of technical and leadership training opportunities. Many federal ministries have limited-term research contracts, as well as annual recruitment fairs. "Geography" or "geographer" may not be mentioned in the advertised job titles, so be sure to read the job description to determine whether the skill set fits your own. Doing some research with the assistance of a guidance counselor or university professor may help determine what opportunities may exist, whether in the nation's capital city, at home where you live, or elsewhere in the world.

PROFILE 7.3

Sarah Brinegar, Ph.D., Social Science Analyst, U.S. Department of Justice (Washington, DC)

Throughout her career, Sarah Brinegar has been interested in issues confronting disadvantaged populations in the United States. Having been a "nontraditional" geography student—a single mother of three from a working-class background—Sarah wanted a job that would not only provide financial security but would give her "the chance to do something meaningful." While her position as a tenured Associate Professor at Marshall University, in Huntington, West Virginia, met these criteria, she found her options within academia to be somewhat limited in terms of promotion and compensation.

In 2008, Sarah changed careers, becoming a Social Science Analyst in the U.S. Department of Justice's Civil Rights Division (Voting Section). In the Voting Section, Sarah provides valuable assistance to attorneys as they work together to enforce the Voting Rights Act. For jurisdictions covered by Section 5 of the Voting Rights Act, for example, she and her colleagues evaluate redistricting plans to ensure that they would not diminish minority voters' ability to elect candidates of their choice. Sarah also helps the Voting Section team evaluate whether racial vote dilution is occurring in a jurisdiction. She typically looks at whether one of the U.S. Supreme Court's preconditions to pursuing Section 2 voting rights litigation is met. This project involves using GIS to draft an "illustrative plan" to see if at least one majority-minority electoral district can be drawn. "It's wonderful to watch cases that I helped start by drawing districts result in voters being able to elect representatives of their choice," she explains.

In addition to providing a superior salary and benefits package and access to the urban amenities of Washington, DC working for the DOJ has afforded Sarah a degree of autonomy comparable to that she experienced as a faculty member. When she's not reviewing or creating illustrative plans, her projects are highly self-directed. She draws on her teaching experience to train new staff in political and population geography concepts and in research skills such as accessing U.S. Census data. In all of these tasks, Sarah relies heavily on her academic background, both in terms of subject matter drawn from her training in social, urban, and human geography and transferable skills such as quantitative methods and writing. In particular, she has found that the versatility she developed as a geographer is a critical aspect of her work, which is highly interdisciplinary. (Her immediate team members are a statistician and a historian.)

As in other branches of the federal government, the Department of Justice offers many opportunities for geographers. The expansion of geospatial technologies means that even the attorneys in Sarah's section are learning the basics of ArcGIS these days. Staff with a strong geographic perspective are more important than ever because they ensure that GIS is being applied effectively and appropriately. Sarah recommends that students strive to become as well rounded as possible and that all geographers emphasize the integrative aspects of their training when meeting with prospective employers. "You need to know how to talk to nongeographers," she advises. "They're not going to know they're looking for you; you have to tell them. You really have to internalize [your interdisciplinarity] so you can verbalize it, so you can sell yourself."

Sarah feels that it's important to embrace reasonable risks and seek new challenges; her philosophy is "You can't be comfortable where you are for too long." She favors candidates who are open to growing in the job, who keep an open mind, and who are willing to learn. She points out that the federal government actively seeks to build a diverse workforce, particularly in the DOJ's Civil Rights Division, so opportunities are available for applicants from a wide range of backgrounds. Sarah's message to prospective employees is: "If you have the skills and want to make a difference, then apply."

—JOY ADAMS

The Canadian federal government has a postsecondary recruitment program that provides a variety of entry-level positions. These positions are broken down into a number of career streams (such as business and social services, research, and program delivery and support). This program is very competitive and recruits on an annual basis, with very specific requirements (Public Service Commission of Canada 2011). It and similar programs provide geographers with a challenging and supportive transition into the public sector. As a student, start by looking at the requirements of programs such as these and begin building your experience now through formal and informal opportunities.

There are numerous informal ways to gain marketable experience and stand out from the crowd. Volunteerism is a valuable way of gaining skills and building networks in your community. It demonstrates commitment to potential employers, builds marketable skills (such as leadership), and may provide opportunities to use the skills and knowledge you have gained in your studies. Voluntary roles in the community may also spur a passion in a specific area, such as social policy or environmental resource management. Get involved in your school, community, and professional geographical associations that offer opportunities to learn more about your field by attending conferences. Also look for opportunities to apply your skills through campus bulletins. In addition, professors or postgraduate students may need students to help with research.

Other informal opportunities for thinking and planning ahead may include reaching out to people in the career area you want to enter. Professionals are often more than willing to talk about their roles if you engage them in an appropriate manner. If you have specific questions you want to explore, reach out through e-mail or, even better, informal interviews over coffee. If you come prepared with specific questions and respect people's time, a short exploratory conversation can yield very valuable information.

Geography is a rapidly growing, interdisciplinary field that has been transformed by the development of GIS and the use of this remarkable technology in an expanding number of applications. As we have attempted to demonstrate using both American and Canadian examples, geographers and the work they do are increasingly valuable to federal government agencies. We hope we have given you an appreciation of some of the career opportunities available in which you can use your geography abilities for the betterment of communities, environments, and society as a whole.

REFERENCES

A Community on Ecosystem Services. http://conference.ifas.ufl.edu/ACES/ (accessed 14 January 2011).

Banks, G. 2009. *Evidence-based policy making: What is it? How do we get it?* ANU Public Lecture Series, Australian and New Zealand School of Government (ANZSOG)/Australian National University (ANU). 4 February 2009, Canberra, Australia. http://www.pc.gov.au/speeches/cs20090204 (accessed 7 November 2010).

Bureau of Labor Statistics, Department of Labor. 2011. http://www.bls.gov/oco/cg/cgs041.htm (accessed June 2011).

Canadian Association of Geographers (CAG). 2009. The origins of the Canadian Association of Geographers. http://www.cag-acg.ca/en/origins_of_cag.html (last accessed 20 May 2011).

Canadian Environmental Assessment Agency. 2011. Whistler Nordic Centre: better through environmental assessment. http://www.ceaa-acee.gc.ca/default.asp?lang=En&n=BD279D10-1 (last accessed 10 May 2011).

COGO. 2010. Letter to The Honorable Harry Reid and the Honorable John Boehner. In (p. 3). Washington, DC: Coalition of Geospatial Organizations (COGO).

DOLETA. 2010. Geospatial Technology Competency Model (GTCM). In (p. 28). Washington, DC: United

States Department of Labor Employment and Training Administration.

Federal Geographic Data Committee. 2006. Annual Report. http://www.fgdc.gov/library/whitepapers-reports/annual%20reports/2006-report/html/index_html (last accessed 10 June 2011).

GITA. 2005. What is GIS: A profession, niche, or tool? Geospatial Information and Technology Association (GITA) White Paper.

GITA. 2008. About the technology. Geospatial Information and Technology Association (GITA).

James, A. M. Gray, R. Martin, and P. Plummer. 2004. (Expanding) the role of geography in public policy. *Environment and Planning A* 36 (11): 1901–1905.

Kruse, J. 2010. Phone interview by author, 22 December 2010.

Long, L. A. 2011. Putting the power of GEOINT in your hands. *Pathfinder: The Geospatial Intelligence Magazine* 9 (1): 2.

Natural Resources Canada. 2008. Geological Survey of Canada. http://gsc.nrcan.gc.ca/index_e.php (last accessed 13 January 2011).

Natural Resources Canada. 2011. GeoConnections: Mapping the future together online. Government of Canada. http://www.geoconnections.org/en/index.html (last accessed 20 May 2011).

Pattison, M. 2010. Phone interview by author, 21 December 2010.

Public Safety Canada. 2011. Emergency management. http://www.publicsafety.gc.ca/prg/em/index-eng.aspx (last accessed 7 May 2010).

Public Service Commission of Canada. 2011. Post secondary recruitment. http://jobs-emplois.gc.ca/psr-rp/index-eng.htm (last accessed10 July 2011).

Public Service Commission of Canada. 2008. Facts and figures about the public service of Canada. http://jobs-emplois.gc.ca/centres/presentation/facts-faits-eng.htm (last accessed 10 July 2011).

Shapiro, C. 2010. Phone interview by author, 17 December 2010.

Taylor, L. 2009. Mapquest: Office of the Geographer makes information visual. *State Magazine* 532: 30–33.

8

Geography Careers in Large Businesses and Corporations

Amy J. Blatt and Michael F. Ziolkowski

Many college students are troubled by the question, "How will I find a job after graduation?" The private sector (small, medium, and large businesses and corporations) contains many dynamic and exciting employment opportunities for geographers. Whether you are a student seeking your first job or are considering a career transition into the private sector, our aim in this chapter is to convey what work is like for many geographers employed in the private sector, with a focus on large businesses and corporations.

Thanks to the synthetic and integrative nature of geography, a plethora of opportunities is available for geographers in corporations, especially for those candidates who are flexible and able to adapt to the needs of the company. Surveys and interviews with corporate employers indicate that among the most valued geography skills are those related to spatial thinking and geospatial technology, cartography, and economic geography (Solem, Cheung, and Schlemper 2008). These same corporate employers also need geographers who are skilled in areas such as project management, statistics, and teamwork, and who can conduct themselves in accordance with practical codes of ethics. Working in the corporate sector also requires individuals to self-assess their performance and be prepared to adapt to rapid changes in the workplace and profession (AAC&U 2010) (see Chapter 2 in this book for additional information about the characteristics of working in a corporate environment).

Traditionally, geography-related jobs in large businesses and corporations can be found in the following areas:

- Environmental consulting (e.g., as a geologist, ecologist, sustainability manager, or geographic information systems specialist)
- Software development and database management (as many information technologies now have a geographical or spatial component, such as Google, MapQuest, and Oracle)
- Real estate and retail markets (such as gas companies and grocery store chains, which require location analysis capabilities)

- Large mapping organizations (e.g., the American Automobile Association, Rand-McNally, and the National Geographic Society)
- Transportation and logistics departments of major corporations (including Ingram Micro, Cisco, Hewlett-Packard, and Intuit).

You can find other career opportunities in the private sector by being creative about how your geographic knowledge can enhance a company's operations. In this chapter, we will offer ideas about how you can persuasively present your geography abilities to corporate employers during the interview process. We will also cover some of the general skills you can expect to perform when working for a corporation.

Corporations need to expand their operations around the world to compete, and to do so, knowledge about space at multiple scales is necessary. Analyzing and understanding the impact of differences and interactions among cultural, political, legal, business, human, and natural resources are some of the activities that attract geographers to professional positions with modern global corporations. Geographers are especially suited for this role because of their abilities and expertise with regard to spatial and place-based analysis, cultural differences and commonalities, and scientific synthesis.

One of the greatest values of a geography education lies in the discipline's integrative nature and the ability of geographers to work across disciplinary boundaries, sharing their spatial perspectives as easily with physical and environmental scientists as with social scientists. Despite the breadth of the discipline, the general public often does not fully understand the practical nature of the field of geography. As such, you will likely need to explain your abilities to potential employers. Below, we present four examples to illustrate the value of geography in corporate organizations. We hope that this discussion will encourage you to think "outside of the box" when brainstorming about your first, or next, dream job.

8.1 Logisticians and Supply Chain Analysts: Connecting and Optimizing Operations

Logisticians are responsible for coordinating a product in the supply chain from origin to customer and points in between. Moving goods from their point of production to their end user requires transportation (Figure 8.1). Reverse logistics may require transportation from the end user through a whole new supply chain. The magnitude, movement, and interaction of the goods in transit with the environment are all concerns of the logistician.

Geographers have long recognized that logistics and supply chain management are central to the function and form of economic space (Cowen 2010). One of the best preparations for a career in logistics is a good foundation in geography and its methods for analyzing and visualizing the distribution of social and environmental phenomena. Cartographic skills and visualization techniques can simply and effectively communicate large amounts of information to a corporate organization. Geovisualization allows us to detect patterns we otherwise may not have anticipated; geovisualization also reveals problems with bad data, helps us understand large- and small-scale features, and facilitates the formation of hypotheses (Ware 2004).

Most of the standard definitions of geography relate to the discipline's core mission to organize knowledge about the distribution and structure of phenomena over the earth's surface. Logisticians and supply chain managers have a similar mission. Because logistics requires

FIGURE 8.1 Total Combined Truck Freight Flows, United States. *Source:* Office of Freight Management and Operations. ops.fhwa.dot.gov/freight/Memphis/appendix_materials/lambert.ppt

the coordination of various actors and synchronization of the material required to perform business functions, logisticians need to understand the importance of networks and nodes, as well as how and why places are embedded and interact within larger systems. Continuously optimizing and maintaining these logistical networks and supply chains for operational effectiveness, efficiency, and customer service is required to sustain corporate competitive advantage. The accompanying profile of Karl Finkbeiner highlights the practical value of geography in the area of logistics.

8.2 Sustainability Managers: Quality of Life for Future Generations

The adverse impacts of human activity on the natural environment are widely known. In response, many corporations have begun to implement sustainability programs and are hiring managers to administer them. When, for example, Frito-Lay, a division of PepsiCo, wanted to create a sustainable potato chip plant in Arizona, it had to develop novel methods of using less water and power while reducing waste (Martin 2007). As more firms try to reduce their greenhouse gas emissions and seek ways to conserve energy, they will need people whose skills transcend traditional academic boundaries and who can bundle different types of technologies into one solution.

Sustainability managers are also responsible for gathering and analyzing data, setting goals, planning and developing tactical solutions to fulfill the corporate sustainability strategy, engaging stakeholders, and continuously improving and verifying the effectiveness of sustainability programs. They often hold high-profile positions with the potential to boost their companies' public image. Geographers, especially those with business backgrounds, are well positioned for these types of jobs because of their broad training across the arts and sciences. Check out Andrew Telfer's profile to see how a merger of geography and business expertise allowed him to pursue an environmentally oriented career in the private sector.

PROFILE 8.1

Karl Finkbeiner, Operations Analyst, Ingram Micro (Buffalo, New York)

After five years as a professional ski instructor in Jackson Hole, Wyoming, Karl Finkbeiner looked to develop new career opportunities in the field of international business. Karl was a history major, but minored in geography as an undergraduate at St. Lawrence University. Because he liked the **real-world, interdisciplinary nature** of the field, he looked for graduate programs in international trade and geography. Drawn to the Geography Department's international trade concentration and its applied focus, Karl went on to earn a master's and a Ph.D. from the University at Buffalo. Today he works at Ingram Micro, the world's largest technology distributor, recently ranked 75th in revenue by *Fortune* magazine.

On a typical workday, Karl deals with various aspects of the company's transportation and logistics operations—working, for example, on the transportation management system to ensure orders are shipped correctly and freight financials are accounted for accurately. While many of his day-to-day duties are not explicitly geographic, he notes that his work "builds on the **basic research skills** of geographic analysis." He explains, "Everything we do here is part of a methodical process, something for which geographic training is very useful."

Unlike many geographers who earn a doctoral degree, Karl decided to hold off on an academic career in favor of work in the private sector. He got his "foot in the door" by networking with a former University at Buffalo geography alumnus who was hiring for a project to monitor and track up to 14,000 shipments per day as part of a loss prevention initiative. Pretty soon, he was employed full time at Ingram Micro, working primarily in freight audit and payment. Karl notes that one of the main benefits of employment in the private sector is the ability to observe measurable progress and the impacts of change on various processes, which contrasts with the less structured work environment of academia. He explains, "The problems that I work on everyday are defined for me, and I know they are valuable to the company."

Karl emphasizes that the **quantitative skills** he learned both as a geography student in lab-based classes and in advanced computer and engineering courses taken outside the discipline are particularly relevant to his job. He adds that **technical skills** such as scripting and managing large databases are also in high demand. According to Karl, "the company thrives on data," and higher level computing skills provide a leg up in a value-added business culture.

Karl believes that the occupational outlook for transportation, logistics, and supply chain management is great because "there's always going to be a desire to move product" from one location to another. The continued globalization of the economy is likely to increase the demand for logistics professionals even further. For anyone considering a career in his field, Karl recommends gaining **practical experience** through internships or other work opportunities. He also believes geography graduates should embrace the flexibility of their education and the multiple opportunities it affords, as many workers today are becoming too "specialized." Karl notes that the average worker today now has many careers throughout his or her lifetime and may perform a variety of functions within the same position. Such trends suit the multidisciplinary nature of geography well: "If you're someone who can synthesize knowledge, you're going to be able to identify and solve the problems you encounter in business."

—MARK REVELL

PROFILE 8.2

Andrew Telfer, Sustainability Manager, Walmart Canada (Toronto, Ontario)

As an idealistic 20-something aspiring to a career in environmental consulting, Andrew Telfer never imagined he'd someday work for the world's largest retailer. But looking back on his career path, Andrew feels "the stars aligned" to bring him to his current post as Sustainability Manager for Walmart Canada.

After completing his bachelor's degree in environmental geography at the University of British Columbia, personal circumstances took Andrew back to Ontario, where environmental career opportunities were relatively scarce in 1994. On the advice of his sister, a human resources professional, he looked for work in sales to obtain **transferable skills** that would enhance his future employability. Andrew quips that his first position, with General Mills, was inherently—if not apparently—geographic, as he was "selling cereal over time and space." However, it soon led to a position with the A.C. Nielsen Company where he engaged his "inner analytical geographer" to work with data on market share, growth, and penetration in the consumer packaged goods industry. When Walmart Canada, one of Andrew's clients, launched its Supercenters, the company recruited him for the position of National Pricing Manager. Shortly thereafter, he ran across an internal posting for a Manager of Sustainability, which he saw as an ideal opportunity to integrate his career experience and academic training.

Today, Andrew observes how closely his work reflects and draws on **geography's broad, integrative perspective**: "Sustainability is much like geography—it's very varied, especially when it's applied to businesses and corporations. It takes the form of packaging, waste diversion, energy conservation, better and more environmentally friendly products for our customers as well for the company's use, and ethical sourcing." As Sustainability Manager, he is involved in every aspect of Walmart's environmental strategy, and, like many geographers, he describes himself as "not a master of any one thing, but a jack of all trades." This diversity equips him to lead a team of experts in specific dimensions of sustainable business practices while he takes the "big picture" view, ensuring that their activities are aligned with corporate goals. Andrew sees sustainability as "the new normal, the new cost of doing business." But this cost can also positively impact a company's bottom line: Walmart's sustainability initiatives in Canada are projected to generate savings of $140 million over the next five years (Gerlsbeck 2010).

With more and more companies "jumping on the sustainability bandwagon," environmentally oriented job opportunities abound throughout the business sector. When evaluating candidates, the main qualification Andrew looks for is **engagement**: "What's on the page is important, but I can teach them the Walmart way." His ideal employee is someone who will "be a sponge and learn" because "no degree program can prepare you for the company you work for." He also points out that **networking is key**, noting that every position he got was facilitated through introductions and contacts.

Andrew's career path demonstrates that the flexibility of training in geography can lead to unanticipated destinations. With persistence and openness to the opportunities that presented themselves along the way, he has come full circle to a career in which he's supporting environmental sustainability at national and global scales. While Andrew Telfer might not have landed the job he dreamed of, he ended up landing the job of his dreams.

—JOY ADAMS

8.3 Geodemographers and Market Research Analysts: It's All about the Customer

Integrating the disciplines of demography, geography, and sociology, geodemography is the science of analyzing people based on where they live (Tonks 1999). Geodemographers seek to understand the processes by which human settlements evolve and neighborhoods form. In a business context, geodemographers map consumer services to "ideal" populations based on their lifestyle and locational characteristics. These profiles are built from geographical databases, electoral lists, and credit agencies to create a picture of population characteristics in different locations. Market research analysts then use the profiles to direct products or services to targeted customers who may be more receptive to those items.

American businesses spend countless dollars perfecting their products and services, upgrading their manufacturing lines, refining their distribution systems, and training their sales personnel. However, these same companies spend relatively little time and money determining which customers are most likely to purchase particular products and services (Egan et al. 2009). And so, while millions of dollars are spent on research and development, businesses often leave out the most important component of the business equation—the customer. As a result, many important business decisions—decisions that could have a huge impact on the organization—are often based on little more than personal and professional advice.

Why have companies ignored the customer? Until the advent of the digital information era, it was difficult to obtain detailed information about the customer base—and even more difficult to interpret this information in a useful and profitable way. Thanks to powerful digital tools and technologies such as geographic information systems (GIS), it is now possible to understand consumer behavior patterns more clearly than ever before. Companies are now able to answer a wide range of critical questions about their businesses with a high degree of confidence—questions ranging from the site of new locations and the allocation of marketing budgets to the best plans for future expansion—all based on new and improved knowledge of who and where the customers are. This capability, in turn, can bring a company higher revenue, faster growth, and enhanced value for shareholders.

In addition to courses in spatial statistics and GIS, geography students can prepare themselves for a career in geodemography by taking elective courses in sociology, economics, and marketing. Excellent communication and presentation skills are also important for these positions. Interested students should therefore seek experience giving well-designed presentations in senior or graduate-level seminars, where one is expected to explain a complex topic to an unfamiliar audience.

At this writing, companies that employ geographers as geodemographers and market research analysts include Buxton Company, Nielson, Rite Aid, CVS, Esri, Experian, The Kroger Company, Marketing Systems Group, and Pitney Bowes-MapInfo, among many others. Macy's is another major company that employs geographers; among these is Esther Ofori, a marketing analyst whose work and professional experiences are profiled in this chapter. In addition, many institutions of higher education have performed geodemographic analysis on their enrollment data, such as Lancaster University (Tonks 1999), The Ohio State University (Marble, Mora, and Granados 1997), and the University of Florida (Thrall and Mecoli 2003).

8.4 "Spatial" Statisticians and Data Analysts: More than Just Number-Crunchers

Although the number of job descriptions that explicitly include the term *geographer* is relatively low, more and more employers are recognizing the value of geographic analysis in their everyday operations. For instance, at the Hartford Financial Services Group, a GIS-working group was

PROFILE 8.3

Esther Ofori, Marketing Analyst, Macy's Inc. (New York City)

Like many of us, Esther Ofori stumbled into geography through a prerequisite course. She's been stumbling into new opportunities ever since, thanks to her hard work, determination, and a little luck.

Esther came to the United States from Ghana to attend college. After falling in love with geography as an undergraduate, she decided to remain at SUNY-Binghamton to pursue her master's degree. For her thesis research, she studied the **location strategies** of ethnic small businesses that were forced to relocate when the Bronx Terminal Market was slated for redevelopment. (It's now the site of the nearly 1-million-square-foot Gateway Center shopping mall.) While attending the annual Applied Geography Conference in 2009, she saw a presentation by a member of Macy's Area Research Team, which is responsible for evaluating new store locations and assessing their competitive impact. She was immediately captivated, seeing how she could leverage her field experience into a career as a **retail geographer**. "I had no room for shyness—I had to talk to him [the Macy's team member]," Esther explains. Putting her nerves aside, she introduced herself to the presenter and told him about her thesis project. Before she knew it, she was invited to Cincinnati to interview for a position as a Research Analyst at Macy's headquarters.

Since then, Esther has worked in a few different positions with Macy's, all of which have emphasized aspects of **geodemography**. In 2011, she returned to New York City to join the Marketing Analytics Team. "We are essentially the resident Area Research Team here but with a focus on the customer," she explains. "We do all kinds of analytical work, but a big part of our job is to **predict customer behavior** by measuring their response to a marketing event or promotion. We use various tools for our analysis and having a GIS and statistics background is a huge plus." Esther also relies heavily on **transferable skills** she honed during her academic course work, including oral presentation, public speaking, technical writing, and the ability to interact within diverse teams.

Esther feels her education prepared her well for her current position. But like all new graduates entering the workforce, she needed some on-the-job training to become established in her career path. Most important was improving her understanding of how businesses operate overall. "When you work in an organization, you are not just working on a project or in your group, but as part of a team to reach the organization's goal," she observes. Success in this pursuit requires flexibility and a willingness to learn.

Reflecting on her experience, Esther gives this advice to job seekers: "Don't give up." "It sounds so cheesy," she laughs, "but it's true." She points out that she worked in a few different positions within the company before finding one that feels "custom-made" for her. Although potential employers don't always understand the value of a geographic perspective, they do need people who can translate data from many different platforms, and a spatial component can be critical to these efforts. For those who can make a strong case for themselves and the discipline, "there are a lot of opportunities out there."

—JOY ADAMS

formed to analyze the impact of coastal hazards and flooding on homeowner insurance rates. None of the participants had the job title of "geographer"; however, the corporation's financial interests necessitated the formation of a working group of personnel trained in GIS and spatial data analysis. The Hartford has also used GIS to understand and analyze customer satisfaction and retention rates. The point here is that, if a business need presents itself where geographic knowledge can play a compelling role, most organizations are willing to make the investment to obtain resources (in physical and human terms) to meet their objectives.

A well-trained geographer with strong quantitative and analytical skills will most likely be able to find a job in a large business or corporation to suit his or her interests and skills. Often, opportune positions for geographers are masked as "data analysts," "statisticians," or "sales territory alignment analysts." Data analyst is often a generic job title that companies use when they advertise for positions involving the management and analysis of data. However, different industries use analysts in different ways. For instance, in the financial services industry, a tax software company, such as Vertex in Pennsylvania, can use GIS to map the different tax codes used in its software. In such a company, a data analyst would be needed to prepare the map and data for the software. When applying for such a position, you may need to show examples of how your geographic abilities support particular types of data analysis sought by an employer.

In the life sciences industry, a data analyst may be assigned to epidemiological studies and analyze demographic data by geographical region. In this position, a geographer could call upon his or her geographic and quantitative analysis skills to produce dot-density and choropleth maps of disease outcomes. These job responsibilities reflect broader developments in the field of epidemiology. In this field, geographers are increasingly contributing their expertise on the relationships between space, place, and environmental factors affecting the spread of disease.

A data analyst in a sales and planning department often participates in sales territory alignment projects. In these positions, knowledge of GIS and spatial analysis can greatly assist the development of marketing strategies used to improve a company's sales performance and profit margins. Check out the profile of Armando Boniche for an example of how a geographer's analytical perspective contributed to new marketing strategies for a major newspaper.

APPLYING FOR CORPORATE POSITIONS AND PERFORMING SUCCESSFULLY ON THE JOB

Although there are no magic formulas or prescriptions for obtaining a corporate job, it is important to keep in mind that business employers are interested in employees who are willing to learn and adapt to the corporate culture. Developing general skills such as oral communication while in school is a very good idea, as most corporate positions involve communicating your work to team members and other departmental staff who may be unfamiliar with geographic concepts. In addition, it could prove advantageous to gain other early professional training experiences that will help you when the time comes to apply for jobs. These include study-abroad programs, internships, professional/industry groups, and fellowships, all of which can help expose you to different people, places, and job opportunities while introducing you to diverse workplace environments and organizational cultures.

Try to take advantage of the available research opportunities within your academic department by acquiring experience in problem-solving using geographic tools and concepts. We also recommend that you practice communicating your findings to a wider audience. One way you

PROFILE 8.4

Armando Boniche, Circulation Director, The Miami Herald/El Nuevo Herald (Miami, Florida)

When Armando Boniche became a geography major, he envisioned a different career path than the one for which his newspaper experience had prepared him. But today, as the newly appointed Circulation Director for *The Miami Herald* and its Spanish-language publication *El Nuevo Herald,* Armando is using geography to get the paper to a quarter-million readers each day.

After working on student newspapers from elementary school through high school, Armando enrolled at the University of Miami as an undeclared major. On the advice of an adviser, he took an introductory geography course, and pretty soon he was hooked. Desktop GIS was just emerging at that time, and Armando was especially drawn to the technological aspects of the field. After completing his undergraduate studies, he decided his next logical step was to pursue a master's degree. As a first-generation Cuban American and a Miami native, Armando wrote his thesis on a topic of both personal and scholarly interest: the influence of demographic shifts on growth and change at *El Nuevo Herald.* With an influx of Latino groups joining the city's large Cuban population, the newspaper was expanding its coverage to include more news relevant to Central and South America. As Armando notes, "Putting Fidel Castro on the front page everyday no longer speaks to everyone."

After completing his graduate program at Syracuse, Armando returned to Miami and began looking for traditional geography jobs in urban planning and local government. In the process, he happened upon a posting for a market researcher/GIS analyst position at *The Herald* that was a perfect match for his skills and interests. He began his career conducting data analysis and mapping to support marketing efforts, which rely heavily on demographic information to reach different audiences. Armando then moved into the role of Circulation Sales and Marketing Director, where he used "geotargeting" to "put more thinking behind what we're doing in terms of [customer] acquisition."

Armando credits his academic advisers, Tom Boswell at the University of Miami and Mark Monmonier at Syracuse, with preparing him to bring his geography skills into the workplace. He explains that the **integration of applied and theoretical perspectives** in their classes "made it easy to make the jump from academics to the real world." For Armando, the kind of **intellectual curiosity** these professors encouraged is the key qualification he seeks in a potential hire. Whether for a research or circulation position, he describes a candidate who wants to learn as "golden". "You need to be very comfortable accepting what you know and what you don't know," he says, adding that new employees should embrace the knowledge of colleagues with more experience, be willing to ask questions, and keep an open mind about doing things they might not have expected to do.

Although declining and increasingly fragmented audiences pose new challenges, Armando is quick to point out that newspaper websites are a major factor driving online media consumption, and they continue to provide much of the content that is picked up by other outlets. "The newspaper business is in transition but there are opportunities for a lot of people, especially in geography," he explains. With the use of **geotechnologies** such as geocoding and interactive mapping to enhance traditional reporting, "information is increasingly being tied to place," creating "entirely new opportunities for the business that weren't evident 10 years ago." For geographers who share Armando's passion for journalism, it is, as he points out, an "exciting time" for the industry.

—JOY ADAMS

can do this is by creating a professional portfolio that includes examples of work you created in your academic courses and research assignments. Professional portfolios can include a variety of materials in both print and online formats; they are a great way to illustrate for employers some of your practical abilities as a geographer.

In the interview process, you may be asked about your unusual choice of college major. Your potential colleagues in the corporate world will want to know how your knowledge and skills can be applied to benefit the company's goals, missions, and objectives. It is important to remember that profitability is the driver keeping your employer afloat and able to retain a staff of skilled professionals. Demonstrating an understanding of your future employer's profit motives and of opportunities for cost reduction will show the employer that you intend to be a valued team player, and that you can anticipate future cost-saving measures or profit-making opportunities.

"Clock Speed" is how Fine (1998) described the metabolism of the modern corporate supply chain. Closely related to the corporation's metabolism is the wide breadth of projects in which employees are expected to participate. This means that your employer will likely expect you to assist with projects of all varieties, and not only those related to your core projects and job responsibilities. Indeed, this will be especially true if you have a history of adding value to projects through your academic training in geography. With this in mind, as you prepare your job applications, make a point of sharing your experiences on service-learning projects, provide copies of reports written for your internship supervisor, or describe your experience with a business research project. All of these experiences can help to demonstrate your practical capabilities as a geographer.

On the job, you will also likely discover that writing is an important activity. Academic training prepares students to write detailed and often exhaustive analyses. In the business world, people are generally much more direct and will want to know, succinctly, what practical steps they should take based on your work that will benefit the firm's goals and objectives. Although writing lengthy reports is certainly required of some professionals in the corporate world (e.g., environmental impact statements), more often you will find that reports, contracts, and memorandums of understanding (MOUs) need to be reduced to short summaries. It can be very difficult, and occasionally frustrating, to reduce one's thorough review of a topic to a one-page report with bullet points. Nonetheless, this is an essential skill that you should master. Keep in mind that you are trying to convince someone to do something and that this requires you to understand the audience's way of learning. Distill your ideas to their bare essence and couch them in the nomenclature of finance and accounting, and success will be more likely.

Many employers seek to improve their competitive advantage and business processes through constant analysis of their operations. This can be unsettling, but it is also necessary in a highly competitive environment. Geographers can add value to their organizations by embracing process mapping and other techniques such as Six Sigma in an effort to aid in continuous improvement. Six Sigma was developed by W. Edwards Deming and Joseph M. Juran, and implemented by Bill Smith at Motorola and others, as a methodology to define, measure, analyze, improve, and control business processes. Six Sigma refers to reducing the probability of defects to six standard deviations from process mean (that equates to about 3.4 errors per million opportunities). Making your organization more efficient than the competition will go a long way toward ensuring your firm's continued success, and in doing so, will provide you with job security.

We hope that we have made it clear that many of the job opportunities available to geography students in large corporations are not well defined. Because many employers are unaware of the practical value of geography, you should be prepared to demonstrate how your geographic abilities can contribute to a company's profitability. It will help your employment prospects if you also demonstrate knowledge of a company's products or services, and show a willingness to adapt to the company's culture and work ethic. Ultimately, the key to success in any corporate-sector career—a desire for continuous improvement—must lie in your own personal initiative.

REFERENCES

Association of American Colleges and Universities (AAC&U). 2010. LEAP and Shared Futures Initiatives. Available at: http://www.aacu.org/SharedFutures/global_entury/index.cfm (last accessed November 2010).

Cowen, D. 2010. A geography of logistics: Market authority and the security of supply chains. *Annals of the Association of American Geographers* 100(3): 600.

Egan, M., K. Manfred, I. Bascle, E. Huet, and M. Sharon. 2009. The consumer's voice—Can your company hear it? Center for Consumer Insight Benchmarking 2009. Boston, MA: Boston Consulting Group.

Fine, C. 1998. *Clockspeed: Winning industry control in the age of temporary advantage*. Reading, MA: Perseus Books.

Gerlsbeck, R. 2010. Inside Walmart's green strategy. *Canadian Grocer*. 12 November 2010, http://www.canadiangrocer.com/top-stories/features/getting-green-1238 (last accessed 14 July 2011).

Marble, D. F., V. Mora, and M. Granados. 1997. Applying GIS technology and geodemographics to college and university admissions planning: Some results from The Ohio State University. *1997 ESRI Users Conference Proceedings*. http://proceedings.esri.com/library/userconf/proc97/proc97/to550/pap501/p501.htm (last accessed 14 July 2011).

Martin, A. 2007. In eco-friendly factory, low-guilt potato chips. *New York Times*, 15 November 2007. http://www.nytimes.com/2007/11/15/business/15plant.html (last accessed 14 July 2011).

Solem, M., I. Cheung, and B. Schlemper. 2008. Skills in professional geography: An assessment of workforce needs and expectations. *The Professional Geographer* 60(3): 1–18.

Thrall, G. I., and N. Mecoli. 2003. Spatial analysis, political support, and higher education funding. *GeoSpatial Solutions* 13(7): 44–47.

Tonks, D. 1999. Access to UK higher education, 1991–98: Using geodemographics. *Widening Participation and Lifelong Learning: The Journal of the Institute for Access Studies and the European Access Network* 1(2): 6–15.

Ware, C. 2004. *Information visualization: Perception for design*. 2nd ed. San Francisco, CA: Morgan Kaufmann.

9

Geography and the Nonprofit Sector

Lia D. Shimada and Jeremy Tasch

GEOGRAPHERS OUTSIDE THE "IVORY TOWER"

From the road, the family farm blends into the surrounding semirural landscape of southeastern Pennsylania. But inside, this farm is also a leading nonprofit organization developing innovative strategies to balance urban edge expansion with sustainable food production. On the other side of the Atlantic Ocean, while uprisings in North Africa through Central Asia grab headlines, a nonprofit organization works in Belfast, Northern Ireland, assisting communities that continue to struggle with a legacy of violent conflict. Nonprofit organizations, whether facilitating community transformations or environmental stewardship, promoting the arts or educational opportunities, and operating at local to international levels, invite us—individuals with geography backgrounds—to use our academic training in real-world contexts. For geographers who seek meaningful jobs—for example, positions that focus on social, environmental, or development causes—employment in the nonprofit sector may be an excellent fit.

Yet although geographers often make ideal candidates for positions within the nonprofit sector, they may face hiring challenges not experienced by applicants trained in more familiar academic or vocational fields. In the United States, in particular, there is limited awareness of what geographers do and what geographers can offer (Standish 2004; Rooney et al. 2006). When searching to fill a position in a nonprofit organization, employers may look to those they know best: economists, environmental scientists, political scientists, and sociologists. However, geographers can draw on a variety of skills and approaches to strengthen the opportunities available to them in the nonprofit sector. One trend worth noting is the adoption of geographic information systems (GIS) and related mapping technologies by aid agencies and environmental organizations (to name just two examples). Although geospatial technologies contribute to the rising profile of geography within and outside of academia, we suggest that geographers representing the wider breadth of the discipline will also find that their particular educational training equips them well for the diverse opportunities available in the nonprofit sector.

Due to the wide variety of organizations contained within the nonprofit sector, the terminology related to them can be confusing: *nonprofit* (with or without a hyphen), *nongovernmental, charity*, and *voluntary*, to name a few. To take just one example, usage of the term *nongovernmental organization* (NGO) is imprecise and varies around the world. In Australia, for instance, an NGO can refer to organizations whose scope is either international or domestic, while in the United States the latter would be known as a nonprofit organization. In the United Kingdom, meanwhile, such organizations are called charities, reflecting the impetus behind their aims and identities. For the sake of clarity, in this chapter we will consider nonprofits, NGOs, voluntary organizations, and the like, as constituting the nonprofit sector, which in turn is also known as the voluntary sector or the third sector (see Gordenker and Weiss 1997; Martens 2002).

In the following section we explore the intersection of geography and the nonprofit sector in order to provide a resource for both geographers and potential employers. We suggest some of the ways those with a geography education can find professional positions within this sector. Next we offer a broad overview of the nonprofit landscape and introduce some of the geographic dimensions of work in this area. We then look at geography as an academic discipline and how it lends itself to employment in the nonprofit sector. The following section explores two forms of engagement between students and employers—internships/practicums and research collaborations—that may be particularly beneficial to those seeking a nonprofit career. We then provide practical information for finding a job. We end the chapter with some concluding thoughts drawn from our own experiences of bringing our geographic training to the work of various nonprofit organizations.

NONPROFITS: A GEOGRAPHIC OVERVIEW

Most U.S.-based readers will be familiar with high-profile nonprofit organizations such as the American Red Cross, Amnesty International, and Greenpeace. These organizations, however, are just a small selection from a sector that operates in broad-ranging areas: education and health, culture and the arts, historic preservation and heritage conservation, environment, science, and religion, to name a few. In their various ways, nonprofit organizations advocate for the well-being of individuals, communities, and cumulatively the world. Their work coheres around issues as diverse as environmental stewardship, children's advocacy and welfare, domestic and international development, provision of support for diaspora and exile groups, and promotion of educational opportunities. These organizations may also be secular or faith-based. They may be freestanding, attached to educational bodies (e.g., some think tanks and museums), or, like the American Red Cross, part of larger regional, national, and international federations of parallel organizations. They may deliver work in stable democracies or in places troubled by political instability and social upheaval. Nonprofit organizations throughout the 20th century played strategic roles in influencing public policies ranging from human rights and the environment to the arts and community development (Bass 2007).

Geographers are well suited not only to the breadth of work described above, but also to negotiating the dynamics of service delivery. Nonprofits tend to be funded by a combination of public and private sector grants, contracts, and sponsorships, and thus are positioned at a complex interface between multiple sectors. Geographers trained to engage with conceptual issues of scale, for example, can bring valuable theoretical and strategic insights to this complex professional landscape. As nonprofit organizations frequently straddle multiple sectors, geographers who analogously have had to straddle various academic disciplines may find their educational experiences well suited to the dynamism of the nonprofit environment. The nonprofit organization's liaising role cannot be underestimated, particularly when working with people who may have acrimonious relationships with government and civic powers. For example, while working in Northern Ireland, Lia administered grants for publicly funded peace-building projects. Her ability to deliver work in paramilitary-controlled communities was possible precisely *because* of the nonprofit status of her organization, while her geographic training helped her to balance micro-level details with macro-level strategic conversations across all sectors.

Nonprofits are a crucial component of robust civil societies. Amidst their diversity, they share a common mission to serve public interests that perhaps may otherwise be underserved or underrepresented. They have traditionally been created by individuals in order to achieve a broader common purpose. Even as early as the American Revolutionary War, for example, volunteer-based peace and resistance organizations formed in reaction to military tax collection and to create civil alternatives to armed conflict. By the mid-1800s, one of Alexis de Tocqueville's (1835, 1840) observations was that in contrast to his experiences in Europe, individuals in the United States and Canada had been organizing into active groups and associations in order to promote social interests rather than self-interests. When the U.S. Congress adopted the federal income tax in 1913, nonprofit organizations were distinguished for the social goods they promoted and were granted federal tax exemptions.

The functions of nonprofits, while maintaining their recognizable legacy, have continued to evolve. Toward the end of the 20th century, in the wake of Gorbachev's 1990 New Year's Eve speech and the demise of the Soviet Union, international attention as well as funding increasingly turned toward those nonprofit, nongovernmental organizations that were oriented toward development, social welfare, and the environment. They became particularly important as governments in the United States and United Kingdom continued the state retrenchments that had accelerated under Ronald Reagan and Margaret Thatcher in the 1980s. Environmental stewardship, arguably an activity with the potential to affect the lives of everyone anywhere, has traditionally been a public service activity promoted by the nonprofit sector. A quick glance at The Great Green List website, for example, offers an informative cross section of 13 nonprofit organizations involved in public outreach, research, and policy analysis regarding climate change.[1]

Broadly speaking, then, the purpose of a nonprofit organization is to benefit the public. Unlike profit-driven organizations, nonprofits do not distribute their surplus earnings to owners and shareholders, but rather reinvest them into themselves to further their goals. Furthermore, a nonprofit vests control not in private owners but in a governing board that holds responsibility for ensuring that the organization fulfills its purpose. To maintain their operations, nonprofits tend to rely on external funding such as grants, donations, and government funds, and are generally exempted from paying taxes. Many nonprofits face daunting financial challenges, as they may operate on shoestring budgets and continually seek new opportunities and sources of funding. Complicating the process is the need to balance the budget at the end of each fiscal year so that an organization remains financially solvent but with a slim enough surplus to warrant its status as nonprofit. As we will discuss in later sections of this chapter, geographers who seek employment in this sector can hold a crucial advantage if they can develop strong fund-raising skills to complement their other strengths.

Funding dynamics are therefore a crucial distinction between nonprofit and corporate entities, and they subsequently shape an organization's operational values. Nonprofits tend to subscribe to an ethos of service that, while potentially present in the corporate and government sectors, is especially pronounced. Their activities are typically aligned with the social values of individuals who are committed to working with others for community improvement as variously

[1]For a listing of nonprofits and their contact information, see "The Great Green List" website at: www.greatgreenlist.com/Climate-Change-Non-profits-106.html (accessed 15 July 2011).

envisioned and practiced. Geographers who are drawn to the nonprofit sector may begin their careers because of interest in, or concern about, the causes with which nonprofits engage. Indeed, much geographic research and curricula dovetails with the missions of nonprofits, particularly the cultural, environmental, historical, community, educational, social service, advocacy, political, and professional organizations that are involved in some of today's most complex concerns (Williams 1996; Gersmehl 2005).

An ethos of service is an important part of the culture of the nonprofit sector, particularly when considering the possible difference in salaries for employees of nonprofit organizations and their public and private sector counterparts. For recent graduates in particular, salary and personal satisfaction are two common considerations when contemplating a career path. The specter of student loans, daily expenses, and possibly the need to sustain a family are very real considerations when one is seeking a job. Concurrently, many geographers aspire to a position where they can apply their academic training in a stimulating professional environment. There is a common perception, particularly outside the sector, that a nonprofit position entails sacrificing a healthy salary out of devotion to the cause of their particular organization. It is important to recognize that this perception—that the characteristics of nonprofit organizations are somehow by necessity extended to "nonprofit" employees—is not actually the case within many nonprofit organizations (Hopkins 2001).

While individuals who feel passionate about particular issues do indeed often staff nonprofits, the choice to work within the nonprofit sector does not necessitate a noncompetitive salary. Many nonprofit organizations compete with for-profit organizations for an overlapping pool of applicants. Thus entry-level salaries at well-established nonprofits are generally competitive with those at for-profit corporations and in some well-funded organizations may indeed be surprisingly competitive. For example, midlevel managerial positions with nonprofit foundations such as the Pew Charitable Trusts can approach an annual salary of $100,000 or more, with professional benefits such as travel and conference attendance.

For nonprofit organizations unable to offer strongly competitive salaries, there may be other attractive benefits. Some Washington, D.C.-based nonprofits, for example, may allow employees an option to work from home during the heat of July. International NGOs may offer as much as six weeks' vacation in addition to *both* local and U.S. or UK (depending on head office location) holiday leave. Depending on the field, some nonprofits—like those awarded with contracts by the United States Agency for International Development—may offer generous compensation packages, including housing allowances and paid visits for employees to their home of record.[2]

BRIDGING GEOGRAPHY AND THE NONPROFIT SECTOR

In this section, we reflect on the specific ways in which geography can serve the nonprofit sector. The study of space and place lies at the heart of the discipline, informing the skills that geographers bring to their analyses and engagements. This orientation provides a unique narrative framework for thinking about the relationships between people, place, community,

[2]As a guide to salary ranges for administrative positions, see www.officeteam.com/salarycenter (accessed 15 July 2011). For positions such as director or program manager within the nonprofit sector, frequently updated Web resources such as payscale.com offer salary ranges and comparisons, as well as open positions (see, for example, www.payscale.com/research/US/Job-Executive_Director_Non-Profit_Organization/Salary) (accessed 15 July 2011). A website such as Simplyhired.com can offer salary comparisons for NGO positions. See www.simplyhired.com/a/salary/search/q-NGO) (accessed 15 July 2011).

environment, and society. Conceptually, geographers are trained to make connections between distinct scales and spatialities. Thus, as we alluded in the previous section, geographers can offer valuable skills for facilitating relations between the nonprofit, public, and private sectors. Furthermore, the breadth of study within the discipline can illuminate new perspectives for nonprofit work. For example, in geographical studies of cultures, diverse forms of social differences as expressed through ethnicity, gender, sexuality, religion, power relations, location, and so forth are explored to better appreciate the ways by which people make sense of their world(s). Contemporary urban geography approaches cities as complex human systems that are reconstructed and maintained daily through global interconnections, class relations, and individual and community practices. Environmental geography approaches the earth's surface as constructed by intricate interactions between natural as well as human processes at various temporal and spatial scales. In short, students of geography are creatively and practically positioned to apply their disciplinary perspectives to the nonprofit sector.

During their training, geographers develop a broad and flexible range of skills—particularly by conducting geographic research—that represent an invaluable resource for nonprofit organizations. A cross section includes preparation and experiences in the following:

- analysis (quantitative and qualitative)
- proposal and report writing
- individual initiative and small-group collaboration
- persuasive communication in written, graphic, and oral form
- public speaking and presentations
- cooperative, creative, and flexible learning
- work with diverse individuals and within teams
- negotiation within diverse environments and among individuals
- coordination of simultaneous projects and prioritizing

Although these types of experiences are not exclusive to the discipline, a geography curriculum can offer students varied training in all of these professional skills, ranging from reflective small-group assignments to research requiring synthesis and analysis to the diverse activities for which computer applications such as GIS are used. Sometimes less tangibly summarized by a particular course, employers can be apprised of geographers' abilities to communicate in graphic, written, and oral forms, their training in resolving quantitative and qualitative problems, and their nuanced appreciation for the relationships between people, place, culture, and environment. The profile of Jane Daniels provides an excellent case for how professional training in geography can provide a breadth of preparation for work in this sector.

On a practical level, a geography curriculum can promote worldviews and sensibilities that carry classroom learning into professional organizational cultures. To this end, educators can play important roles in promoting awareness of geographic learning as the basis for professional training. For example, as both an educator and a former director of an international nonprofit organization, Jeremy encourages collaboration, teamwork, small-group activities, and experiential and service knowledge as the basis for an interactive learning and teaching process that is highly transferable to positions in the nonprofit sector. Also integral to a contemporary geographic education are an appreciation and understanding of diversity, combined with skills for negotiating differences, and building communities that respect and acknowledge pluralism and environmental issues. Geography coursework also engages seriously with the difficulties of subjectivity—"our" sources of values and those of "others"—and calls on students to grapple with challenging ethical, moral, and human dilemmas. These educational components are

PROFILE 9.1

Jane Daniels, Director of Preservation Programs, Colorado Preservation, Inc. (Denver, Colorado)

While growing up as a first-generation Czech American in an ethnic community near Milwaukee, Jane Daniels developed a keen interest in her own heritage as well as the cultural backgrounds and customs of others. However, she didn't consider heritage as a potential career path until she took a graduate course in historic preservation at the University of Wyoming while pursuing her master's degree in international studies and environmental conservation. "I instantly saw the overlap between heritage and my undergraduate training in geography and environmental studies," she says. "I saw it as an opportunity for doing something that was meaningful to me and that I had been trained in."

Jane's undergraduate education at the University of Wisconsin-Madison was excellent preparation for her position with Colorado Preservation, Inc., a private, nonprofit statewide historic preservation organization. "As geographers, we have a tendency to look at the world in a broad sense," she explains. Preparing a proposal explaining why a site should be preserved and developing plans for how it should be interpreted require her to **analyze a wide array of data**. Jane needs to consider the site's historical, cultural, political, and economic contexts as well as material aspects of the surrounding landscape, such as transportation networks, zoning restrictions, and local architecture. She also needs to carefully evaluate the costs, benefits, opportunities, and limitations of each potential project. In her opinion, professionals with an **interdisciplinary background** that combines the liberal arts and the natural sciences are particularly well equipped for this sort of work.

Jane's position as Director of Preservation Programs encompasses a number of additional responsibilities, including developing funding strategies, financial reporting and budgeting, public relations, and hiring contractors. Cooperating with a broad, diverse group of stakeholders is a key skill, which requires well-developed abilities in oral communication, writing, and public speaking. Jane also got **on-the-job training** in several important areas that weren't included in her university curricula, specifically real estate, financial management, and architecture and construction. She strongly recommends that aspiring preservationists seek **hands-on experience** in the field. Before joining Colorado Preservation, Inc., in 2008, Jane was the Executive Director of the Main Street Program in Laramie, Wyoming, and she previously worked in the private sector as a consultant to rural communities and in the public sector for the city of Madison.

Jane's ideal job candidates also demonstrate **consistency in their interests** and a commitment to pursuing them. On working in the nonprofit sector, she observes: "There's an opportunity to become more personally invested in whatever work you're doing. There's a chance that you'll fall in love with the buildings and projects you work with, so we tend to work very hard and become passionate." Volunteer work provides evidence of one's dedication to the cause as well as tangible experience in various aspects of historic preservation. "There are many opportunities to get involved without applying for a job outright," Jane reports.

The current housing slump has slowed real estate markets, but this situation has opened doors for historic preservation. With new construction down, there is increased demand for the preservation and restoration of existing structures. Furthermore, growing interest in sustainable development, economic revitalization, and livable communities all have positive implications for the field, which is growing and gaining recognition. As historic properties are increasingly acknowledged as worth preserving, there is more room for people in the field who are specialists like Jane as well as for those who may not be experts just yet, but who can offer relevant knowledge and skills.

—JOY ADAMS

directly transferable to professional work in nonprofit organizations. Employment within a nonprofit, either domestically or internationally, commonly involves interacting with diverse members of a particular community such as youth and the elderly, government and business officials, students and educators, as well as with other nonprofit representatives. Navigating these diverse and overlapping groups is part of the challenge and the excitement offered by nonprofit employment.

Although job prospects in the nonprofit sector increase considerably for degree holders, we suggest that the undergraduate and graduate training processes can be harnessed intentionally for bridging educational development and professional experience. In the next section, we describe some entry points for this type of engagement.

FINDING "THE DOOR"

Even before they receive their diplomas, undergraduate and graduate students offer tangible benefits to organizations in the nonprofit sector. Volunteer positions, internships, and practicums can introduce geography students to a sample of the types of work and organizations available upon graduation. In general, an organization can offer a formal role to a student who either volunteers or agrees to work for a modest stipend. These introductory positions provide opportunities to develop networks, relevant skills, and invaluable professional experience. Frequently, these positions present an insider's view on upcoming openings for full-time jobs. The organization, in turn, benefits in the short term from the students' labor, and in the longer term by investing in the development of professionals who will further its missions, values, and aims. During his time as a director of an international educational nonprofit organization working in Azerbaijan, Jeremy created a program to bring students into internship programs with the Ministries of Ecology, Education, International Affairs, and Communication and Information Management. To date, fifty-five Azeri students have completed internships, and several are now full-time government employees.

Although many organizations, like Jeremy's, offer formal internship programs, the idea may not have occurred to other nonprofits, particularly smaller organizations with limited budgets and staff capacity. For these organizations, a student may wish to take the initiative by establishing contact and offering his or her services. An important consideration to bear in mind is that managing a volunteer requires an organization to commit sizeable human resources. To this end, the student can take certain steps to facilitate the process while making the prospect of an internship more attractive to a potential employer.

Before approaching an organization, the student should take time to consider the scope of his or her role within the organization and how it can be managed. The student will wish to emphasize the ways in which he or she is prepared to work independently, with limited direct supervision. Above all, the student should present clearly how this role will enhance the work of the organization, particularly in relation to his or her training as a geographer. Indeed, for employers who are further removed from education, a student of geography can bring particularly fresh insights to an organization. For example, Jeremy found that combining classroom pedagogy and scholarly practice worked well in the professional setting of an international nonprofit. Working as a team, his interns and colleagues hosted a major international conference that brought together government, corporate, and nonprofit representatives to discuss interregional policy in novel ways. The nonhierarchical collaborative discussions, though familiar to students of geography, represented a new approach for these diverse conference participants and the organizations they represented.

In addition to internships and practicums, research collaborations can be another powerful form of engagement for geographers who seek to work in the nonprofit sector. Research collaborations also have the benefit of contributing to rich methodological dialogues within the discipline. In recent years, "participatory geographies" have reached a critical mass (for example, see Cahill 2007; Kesby 2007; Kindon and Kesby 2007; Pain and Kindon 2007). Broadly speaking, participatory research refers to approaches in which the conventional subjects of research are involved in the research process itself—from articulation of questions and design of methodologies to dissemination and action. Through this process, the participants and the academic researcher(s) share ownership of the project (Pain 2003). Kesby (2007, 2813, citing Pain 2004) observes:

> Participatory research is well suited to social geography and to the local scale at which much qualitative fieldwork is conducted. Its many innovative techniques can revitalise geographical methodology and offer new opportunities for the perspectives of the marginalised to emerge. Participatory approaches also aspire to a broader notion of ethical research than the conventional wisdom of "do no harm": by creating new spaces for critical engagement beyond the academy, they facilitate arenas in which participants and researchers can collaboratively generate knowledge and informed action.

By way of example, we draw on Lia's successful experience of gaining employment through the participatory research process. Her doctoral dissertation, which focused on cultural geographies of peace-building in Northern Ireland, emerged from a longstanding interest in participatory geographical methodologies. Given the precarious political situation in Belfast, she chose to collaborate with Groundwork Northern Ireland, a community regeneration nonprofit that has strong relationships with paramilitary-controlled neighborhoods on both sides of the sectarian divide. Working under an inherently geographical motto—"Changing Places, Changing Lives, Changing Minds"—Groundwork emphasizes the social regeneration of these post-conflict communities. During the course of her fieldwork, her relationship with the organization shifted dramatically, from an initial role as a volunteer to formal employment as a project officer. Lia's research experience can be read as a process by which, through her professional responsibilities, she staged the encounters she wanted to study—for example, the intersection between racism and paramilitarism in Northern Ireland.

As participatory research is inherently collaborative, nonprofit organizations may present a logical choice for partnership. Whether they support the arts, provide educational opportunities, contribute to public policy formation, or promote environmental stewardship, nonprofits generally and collectively promote participatory democratic change and enhancement. For participatory involvement to occur, learning, research, collaboration, and dedication all must be intertwined. Thus formal and informal employment within the nonprofit sector is particularly appropriate for geographers who consider civic engagement a professional and academic priority.

GETTING YOUR "FOOT IN THE DOOR"

Unlike management and business graduates, who are trained to think in terms of professional application, students of geography may not realize that their own academic training offers highly transferable skills for a competitive and varied job market.While employment solicitations for

PROFILE 9.2

Kate Pearson, Strategic Partnerships Director, Habitat for Humanity International (Port-au-Prince, Haiti)

Working in Haiti has taken her far from her home state of Alaska, but Kate Pearson's job allows her to address issues that are close to her heart. "Even the poorest people living in the U.S. have it good compared to Haiti," she observes. "That harsh reality keeps me going."

As a child, Kate thought "old maps were really boring." But when it came to choosing a career path, the apple didn't fall far from the tree. Her father is a geography professor, her mother is a historical cartographer, and several other family members are geography or social studies teachers. After a semester in Ecuador piqued her interest in international poverty and inequality, Kate pursued a bachelor's degree in geography and environmental studies at Middlebury College and a master's degree in geography at the University of Arizona.

Kate joined Habitat for Humanity International's Haiti operation in February 2010, after five years at the organization's Latin America and Caribbean Regional Office. Her primary responsibility is to develop and maintain partnerships with donors, including bilateral institutions, other NGOs, corporations, and faith-based groups. Prior to the devastating January 2010 earthquake, Habitat had provided housing solutions to more than 2,000 Haitian families. To date, it has helped improve conditions for 30,000 families in communities affected by the disaster, which destroyed or damaged roughly 200,000 homes. To support these efforts, Kate and her team have raised over US$45 million for disaster response programming.

Natural disasters are mostly human disasters, especially in settings like Haiti, where so many structures collapsed because they weren't built to withstand earthquakes. Rebuilding safer homes reduces vulnerability to future hazards while creating jobs and training opportunities for residents and boosting local economies. Efforts to understand the connections between housing and concerns such as public health, water supply and sanitation, environmental quality, safety and security, and cultural and gender sensitivities mean that geographers and other professionals with interdisciplinary backgrounds have valuable perspectives to contribute to these projects.

Kate notes that regional foci can sometimes create silos within international development organizations, so she advises prospective employees to acquire **broadly transferable skills**, such as grant writing, communication, basic mapping and GIS proficiencies, foreign languages, and project management, rather than narrowly focusing on a particular region. While specialized skills and knowledge can be an asset, she would usually rather hire someone with **hands-on experience** and no master's degree than vice versa. For example, international fieldwork in challenging environments provides critical preparation for the difficult conditions often encountered in developing regions.

Because the field needs people who are **proactive self-starters**, Kate encourages job seekers to focus on highlighting their accomplishments and to do research on the organization and its mission in advance of making contact with a prospective employer: "You want to show that you have a mindset of 'getting things done.'" Kate estimates that a majority of positions today are filled through face-to-face contact, so she recommends digging into contacts within your networks, internships, and informational interviews as strategies for getting in the door.

"International development is a professional career, and it requires all kinds of professionals," Kate observes. "It's a microcosm of the world as a whole in terms of opportunities." Geographers can contribute their broad skill sets and the **cross-sectional, holistic approach** central to many development projects. Kate's work is intrinsically rewarding and meaningful despite the day-to-day challenges she faces. For her, the job is not only about helping others but "transforming your own life and perspective."

—JOY ADAMS

domestic or international positions with nonprofit organizations rarely specify a geography degree among the list of qualifications, this should not stop a geographer from applying. The challenge is to convince a nonprofit employer that training in geography is relevant to their mission and goals. In her profile, Kate Pearson highlights some of the geographic skills and other abilities that have proven helpful in her position as Strategic Partnerships Director for Habitat for Humanity International.

As you contemplate entering the nonprofit sector, consider the type of organization with which you would like to be associated. Nonprofits may employ only a few individuals or as many as would a small college. Focus your attention on the type of organization, its major emphases, and the nature of its activities (e.g., research, advocacy, direct service, working indoors or out, policy development, or administrative work; see Table 9.1). This point may seem obvious, but the diversity of nonprofit positions can be overwhelming, particularly for geographers with wide-ranging interests. By being selective from the start, you will be better able to create clearer correspondence between your interests and skills, geographic training, cover letter, and curriculum vitae. In this way, you will stand out to a prospective employer, who will recognize your interest in *the* particular position rather than simply *a* position, and who may thus be more inclined to invite you to an interview.

It will be vital to consider the needs of the organization and the ways in which you can address these needs. For a nonprofit organization to remain relevant in its field and to attract funding, it must constantly seek ways to diversify its services while staying true to its core values. To this end, you might prepare some ideas for networking, fund-raising, or raising the organization's profile with new audiences. As we mentioned in the second section of this chapter, we encourage you to emphasize your aptitude with scalar and spatial analyses by offering concrete examples of how this way of thinking could benefit your employer. Flexibility and imagination are additional key professional attributes that you can easily transfer from your academic training in geography.

TABLE 9.1 Resources for Employment in the Nonprofit Sector

Name	Description
GuideStar (www2.guidestar.org/AdvancedSearch.aspx)	Serves as an extensive and searchable source of information about and for nonprofits, helpful for identifying individual organizations within specific cities and states.
Idealist (www.idealist.org)	Provides information regarding volunteering, internships, consultancies, and conferences related to the nonprofit sector.
Non Profit Yellow Pages (www.nonprofityellowpages.org/Ypsearch.asp)	Provides a searchable database of vendors and consultants that work within the nonprofit sector.
PNN Online: The Nonprofit News and Information Resource (www.pnnonline.org)	Offers information and a variety of resources regarding the nonprofit sector.
The Riley Guide (www.rileyguide.com)	Provides a directory of employment and career information sources and services.

If you are ready to pursue your interest in working within the nonprofit sector, here are a few guidelines for your job search:

1. ***Assess your background and interests in working with a nonprofit organization.*** You will be a more competitive and desirable candidate if you write and speak persuasively about your desire to work on particular nonprofit issues. Beyond appropriate formal coursework, read up on the concerns relevant to a particular field or region of interest. What do you care about most: the environment; health care; human trafficking; peace negotiation; education; the arts; poverty? As a geographer, do you envision bridging two or more of these areas?
2. ***Find an organization and an open position.*** Through an Internet search (see Tables 9.1 and 9.2 for suggestions), your university's career services or international studies office, or the resources offered in this book, identify several organizations with which you would consider working. The websites of most nonprofit organizations feature an employment link that lists current position openings. In this manner, it is a rather straightforward process to match current opportunities with your interests and experiences. Do enough research on the organization (how they get their funding, what their main programs are, their mission statement) to get a sense of where they are headed, and how you could contribute to their work.
3. ***Seek an informational interview.*** When positions are not listed at your organization of choice, or even if they are, requesting an informational interview is an excellent way to introduce yourself to a potential employer. Individuals already working for a nonprofit may agree to meet you for an informational interview. To this end, your college or university's

TABLE 9.2 Selection of Nongovernmental and Nonprofit Organizations

Organization	Website
Adventist Development and Relief Agency	www.adra.org/site/PageNavigator/about_us
American Councils for International Education	www.americancouncils.org/aboutOrg.php
Catholic Relief Services	www.crs.org/about/
Chemonics International	www.chemonics.com/aboutus/aboutus.asp
Counterpart International	www.counterpart.org/about
Civilian Research Development Foundation	www.crdf.org/join/
DAI	www.dai.com/work/
Eurasia Foundation	www.eurasia.org/about/
Habitat for Humanity	www.habitat.org
Human Rights Watch	www.hrw.org/en/about
International Foundation for Electoral Systems	www.ifes.org/About/Who-We-Are.aspx
International Republican Institute	www.iri.org/learn-more-about-iri-0
International Research and Exchanges Board	www.irex.org/about-us
National Democratic Institute	www.ndi.org/whatwedo
Open Society Institute	www.soros.org/about
Pacific Environment	www.pacificenvironment.org/article.php?id=58
Transparency International	www.transparency.org/contact_us/work
World Wildlife Fund	www.worldwildlife.org/who/

alumni network can be a powerful resource. In an informational interview, you are actually doing the interviewing as you seek information about the organization, while concurrently demonstrating your interest in and preparation for a position. This type of interaction may or may not lead eventually to a job offer, but at a minimum this is an excellent way to develop professional contacts while gaining interview experience. Prepare a range of questions in advance of the meeting. After your conversation, take time to follow up with a polite "thank you" note.

4. ***To secure an overseas position, distinguish yourself from other applicants.*** Larger organizations may have offices in several locations. Turnover may be frequent, and, depending on the type of position, shorter-term contracts of three months to a year may be possible. If working in another country for a nonprofit is of interest, then consider writing a brief letter of introduction to the office director or chief-of-mission in the specific location. Highlight what distinguishes you from other potential applicants and why you are attracted to this particular location. While a local representative office may not be in a position to decide directly on hires, it may be able to influence the hiring decisions made by the human resources department at the organization's headquarters.
5. ***Be flexible, patient, and persistent.*** Acknowledge that securing employment with a nonprofit may require ongoing research, as well as flexibility in position preference, and that some positions include several steps between submitting your application form to receiving an actual job offer. Although the entire process may proceed rapidly, remember that six to nine months can elapse between initially finding and finally securing a position.

Although the process may appear daunting at the beginning, think carefully and strategically about what you can bring to the organization. There is, of course, no secret to securing domestic and international professional positions, but there are particular steps that may help to increase a geographer's opportunities for securing a fulfilling job in the nonprofit sector. The profile of Serge Dedina and Emily Young offers an additional perspective on the traits that many nonprofit employers seek in potential candidates.

CONCLUDING THOUGHTS

When we began writing this chapter, Lia had left her job in Belfast and was in the midst of searching for new employment in the nonprofit sector. Forty applications and eight interviews later, she accepted a position with a large faith-based nonprofit in London that hired her to implement its diversity strategy. This is inherently geographic work that draws spatial analysis, research, and policy development into dialogue with an overarching ethos for inclusion. Over the course of a job search that spanned six months, three continents, and ten cities, Lia found herself grappling at every turn with what it means to be a geographer in a highly competitive market. The experience highlighted the value of academic training that can be invoked in myriad, flexible ways. In particular, she recognized that interviewers tend to look positively on the collaborative geographic research process—from the initiative required to launch a project to the sustained follow-through required to complete it. Lia's research collaboration in Belfast had another surprising outcome: On the organization's behalf, she undertook accredited training in a suite of skills (mediation, principled negotiation, conflict management, and community capacity building) that will serve her throughout her professional life. Indeed, Lia continues to put these skills in practice by volunteering as a mediator for a small, community-based nonprofit in her local corner of London.

PROFILE 9.3

Serge Dedina, Ph.D., Executive Director, WiLDCOAST/COSTASALVAjE
Emily Young, Ph.D., Senior Director, Environment Analysis & Strategy, The San Diego Foundation (Imperial Beach, California)

Emily Young and Serge Dedina met as undergraduate students while studying abroad in Peru, and they were drawn together by a mutual interest in environmental geography and resource conservation in the Baja California region. Since completing their master's degrees at the University of Wisconsin and doctorates at the University of Texas, Emily and Serge have continued their personal and professional journeys together.

Because of their similar academic training and research specializations, Serge and Emily realized it would be a challenge for them both to find faculty positions. Fortunately, Serge found himself particularly drawn to applied geography and decided that a nonprofit career would be a better fit for him. When Emily started a tenure-track position at the University of Arizona, Serge found a job with the Nature Conservancy. After a few years and the birth of their sons Daniel and Israel, they relocated to San Diego. Balancing a young family with the demands of Emily's teaching load and tenure requirements was proving difficult. In addition, they wanted to be closer to Serge's family and to professional opportunities in his areas of interest.

After moving to California, Serge co-founded WiLDCOAST/COSTASALVAjE, a binational organization that works to conserve coastal and marine ecosystems and wildlife (www.wildcoast.net). Emily also transitioned into the nonprofit sector, joining the staff of the San Diego Foundation (www.sdfoundation.org) where she works with community volunteers, donors, and other foundations to direct charitable giving to the region's critical environmental needs. Serge describes his exposure to geographic tools and perspectives as "one of the main reasons I've been successful and WiLDCOAST has been successful." Emily notes that the **analytical framework** she developed as a geographer enables her to understand a variety of regional environmental concerns, ranging from climate change to biodiversity conservation to air and water quality, within their broader social and economic contexts, helping her to determine where funds should be invested in order to have the greatest impact.

Although salaries can be less competitive than those offered in other sectors, nonprofit work is intrinsically rewarding. A **passionate commitment** to the mission of the organization is the key qualification that both Serge and Emily seek in potential employees. **Prior experience** is another important asset. For example, Serge describes his position with the Nature Conservancy as being "like a graduate program for nonprofit management." However, paid employment is only one option for developing job skills; Emily strongly recommends volunteer work and internships as other ways to gain valuable first-hand experience.

Nonprofit employees often wear many hats, so burnout is something of an occupational hazard within the industry. Serge and Emily maintain a healthy work-life balance by integrating family vacations into their professional travel. The photo above shows the family in Oaxaca, Mexico, where Serge recently scoped out potential WiLDCOAST conservation projects and the family observed slash-and-burn agricultural practices and visited neighboring ruins. "It was like being in a cultural ecology field course," he noted, adding that that field research is the "funnest thing you can do as a family." (For more on the subject of work-life balance, see Chapter 14 by Jan Monk in this book.)

"Geography is uniquely positioned to prepare the next generation of well-rounded and innovative thinkers to address environmental issues," Emily observes. The couple recommends that job seekers look to gain experience with the financial and fundraising aspects of nonprofit work, develop the ability to communicate effectively with nonspecialists, and consider pursuing a master's degree for specialized training. "Take advantage of every opportunity," Serge advises. "It's more important than ever that geographers get out there and use their skills and training to make a difference."

—JOY ADAMS

Blurring the dichotomy between disciplinary philosophy and practice, Jeremy began to volunteer with environmental and educational NGOs while conducting graduate fieldwork in Russia. These volunteer initiatives turned into short-term, paid consultancies, which then evolved into several years of international work promoting civil society development from within the nonprofit sector. Now a geography professor, Jeremy continues to volunteer with a local nonprofit and also continues to visit the Caucasus, Central Asia, and Russia, combining research on the human dimensions of environmental change with civil society development, while learning from and sharing with members of the extended nonprofit community.

Whether as a volunteer or a career participant within the nonprofit sector, geographers have the chance to work with others for good causes, to feel inspired by the projects in which they are involved, and to help build better worlds.

Understanding our geographies within and outside academia helps us to understand our worlds of different scales, our places in them, and what roles we can play from students to policy makers in facilitating positive change. Geographers who decide to become involved in the nonprofit sector will come to realize, as have Lia and Jeremy, that a job is a means to earn a salary, that a calling is a way of life, and that work within this sector can be a satisfying way to pursue both.

REFERENCES

Bass, G. D. 2007. Advocacy is not a dirty word. *Chronicle of Philanthropy* 20 (2): 1040676X.

Cahill C. 2007. The personal is political: Developing new subjectivities through participatory action research. *Gender, Place and Culture* 14 (3): 267–92.

de Toqueville, A. Translator: Henry Reeve. 1835, 1840. *Democracy in America*, Volumes 1 and 2. Release Date: 21 January 2006. EBook #816. www.gutenberg.org/files/816/816-h/816-h.htm.

Gersmehl, P. 2005. *Teaching geography.* New York, NY: Guilford Press.

Gordenker, L., and T. G. Weiss. 1997. Devolving responsibilities: A framework for analysing NGOs and services. *Third World Quarterly* 18 (3): 443–55.

Hopkins, B. R. 2001. *Starting and managing a nonprofit organization.* New York, NY: Wiley.

Kesby, M. 2007. Spatialising participatory approaches: The contribution of geography to a mature debate. *Environment and Planning A* 39 (12): 2813–31.

Kindon, S., R. Pain, and M. Kesby (Eds.). 2007. *Participatory action research and approaches: Connecting people, participation and place.* London, UK: Routledge.

Martens, K. 2002. Mission impossible? Defining nongovernmental organizations. *Voluntas: International Journal of Voluntary and Nonprofit Organizations* 13 (3): 271–85.

Pain, R. 2004. Social geography: Participatory research. *Progress in Human Geography* 28 (5): 652–63.

Pain, R. 2003. Social geography, relevance and action. *Progess in Human Geography* 27 (5): 659–76.

Pain, R., and S. Kindon. 2007. Participatory geographies. *Environment and Planning A* 39 (12): 2807–12.

Rooney, P., P. Kneale, B. Gambini, A. Kieffer, B. Vandrasek, and S. Gedye. 2006. Variations in international understandings of employability for geography. *Journal of Geography in Higher Education* 30 (1): 133–45.

Standish, A. 2004. Valuing (adult) geographic knowledge. *Geography* 89 (1): 89–91.

Williams, M. (Ed.). 1996. *Understanding geographical and environmental education.* London, UK: Cassell.

10

Starting a Small Geography Business

Kelsey Brain

"The best reason to start an organization is to make meaning; to create a product or service to make the world a better place."

—Guy Kawasaki, venture capitalist, founding partner of Garage Technology Ventures

The motivation for being an entrepreneur is far more than simply making money. Entrepreneurship is also about contributing to one's community by producing and selling something meaningful. By starting a small business, an entrepreneur has the opportunity to both do something that he or she values and to offer a product or service that solves a problem or provides assistance for someone else. For geographers, starting a business is a chance to contribute to society in a way that few other activities can offer.

Entrepreneurship is thriving in the United States. According to the U.S. Small Business Administration (2010), approximately 600,000 small businesses are founded every year in the United States. These range from businesses run by one person to companies with dozens of employees. And yet despite the popularity of small business ownership, geographers have not traditionally been provided with much training to prepare them for starting a business (Bond 2005; Estaville et al. 2006). Starting a small business means creating a separate and distinct business entity that is formally registered with the state, has a small number of employees, and is usually owned and operated by one or a few people. The business provides services or products to a specific customer base, which may include government agencies, nonprofit organizations, other businesses, or the general public.

This chapter will introduce you to some of the opportunities and challenges of starting a small geography business by reviewing examples of companies founded by geographers. The chapter will also describe some of the nuts and bolts of starting a business, with suggestions for external reading and other resources that can help you.

GEOGRAPHERS AS BUSINESS OWNERS

There are many reasons why a spirit of entrepreneurship motivates geographers to start a small business. Some are attracted to the creative aspects of writing a business plan and developing a product, whereas others may be driven by the inherent challenges and risks of such an undertaking. Being a business owner may also provide the freedom of being one's own boss, the flexibility to be more family-oriented, and the ability to run a business according to one's personal values. Perhaps the most significant reason, which harkens to the introductory quote by Kawasaki, is that entrepreneurs have the ability to help other people and enhance community life by offering needed products and services. And quite possibly, starting a business can ultimately lead to high rates of return on investments and significant earnings.

While you read this chapter, adopt the mindset of an entrepreneur. Be creative and imaginative as you think of how your geography knowledge and skills might be applied to serve a need in your community. When you discover a potential match between a need and your skill set, you may very well have found an idea for a small geography business of your own.

A small business meets a need for a product or service that is not being provided by another company, or it may offer a product at a superior price or quality relative to existing products. Geographers have a unique perspective for solving problems, one that draws on spatial thinking, place analysis, and interdisciplinary approaches. Employing a geographic approach to problem solving, geographers can develop a variety of businesses that serve vital needs of government agencies, businesses, nonprofits, and the general public. Although the possibilities are manifold, here I will highlight three examples of how geographic perspectives have been used to create a small business.

1. A Spatial Perspective of Economic Activity

Geographers examine economic activity within and across space. A primary example of the way geographers study economic processes and patterns is through retail location analysis, often supported by geographic information systems (GIS). By using GIS to analyze spatial data, a small geography business can help its clients determine the best site for a new retail store on the basis of customer demographics, customer proximity, travel routes, and competition. Similar services can be provided in the context of real estate and environmental impact analysis. For example, Kristin Carney founded a small business called CUBIT for this purpose (Carney 2010). Carney worked for an environmental engineer by gathering and analyzing environmental data to determine where a new road should be placed to minimize negative environmental impacts. Because traditional methods of gathering such data were too time consuming, Carney created CUBIT, a program that uses a Google Maps interface to provide compiled environmental data to users within 30 seconds of points being placed on a map. CUBIT reduces the time spent gathering data for a project by environmental engineers and urban planners from over 40 hours to a few minutes (Carney 2010). For more on CUBIT, see the profile of Kristin Carney and her business partner, Anthony Morales.

The example of CUBIT highlights the general need many businesses, research centers, and government agencies have for large volumes of spatial data. Many recent business start-ups have focused on providing similar data. For example, WeoGeo, a veritable iTunes for professional surveyors, engineers, and architects, was founded in Portland, Oregon in 2006 to provide a better solution for map storage, retrieval, and management. The company offers easy-to-use, Web-based mapping and computer-aided design (CAD) content management services that enable developers of professional mapping, surveying, or civil engineering information to share data within their

PROFILE 10.1

Kristen Carney and Anthony Morales, Co-founders, Cubit Planning (Austin, Texas)

A few years ago, Kristen Carney worked in environmental planning. One particular assignment—which she now affectionately calls her "Nightmare Project"—required her to spend hundreds of frustrating hours gathering and formatting data for a road construction project. Her friend Anthony Morales, a Web developer, agreed to help her develop an automated solution using a Google map interface. Instead of spending hours assembling and formatting data for each proposed alternative, their tool allowed Kristen to generate user-friendly tables in just seconds. After receiving rave reviews from fellow planners, Kristen and Anthony realized the potential market for their product and co-founded Cubit Planning in 2009.

As, respectively, an anthropology major and an English major who have built a business on geographic data and maps, Anthony and Kristen are living proof that **careers sometimes arise from unexpected opportunities**. Kristen had always toyed with the idea of starting her own business. In her first job after college, she worked for a self-employed economic consultant who taught her the ins and outs of managing a small business. Anthony describes his father as a "serial entrepreneur." He fought the idea of following in his footsteps "tooth and nail" until Kristen came along with her Nightmare Project; now he says he's "addicted."

Because their formal education didn't prepare them for this career path, Kristen and Anthony's **passion for learning** is essential to their success. "We still learn every single day," Kristen explains. "That's one of the most amazing things about owning your own business, but it can also be overwhelming." Luckily, Austin provides a thriving entrepreneurial community and abundant resources to help support fledging businesses. One example is Capital Factory, a program that has been described as a "ten-week MBA program for technology start-ups." Cubit Planning was selected to participate in 2009, receiving a small amount of seed capital and weekly mentoring sessions by entrepreneurs who have founded successful companies. Running a business requires them to know "a little bit of everything," so the pair also rely heavily on online resources: "Google's our best friend," Kristen quips. Anthony adds that "You no longer have to know everything at a very technical level like you did five or ten years ago," thanks to the abundance of resources and inexpensive "plug and play" solutions that are now readily available.

Having great resources doesn't mean that the job is easy. Anthony is quick to point out that there are few "overnight successes." Kristen notes that exhaustion and money are ever-present concerns, but she adds that they are well worth the satisfaction and excitement that comes with their work. Kristen and Anthony have two key pieces of advice for aspiring entrepreneurs: (1) **Don't do it alone**. If you're a sole proprietor, find a mentor. Read, talk to others, and ask questions. (2) **Start on the side**. Take twenty minutes to write a blog post about your idea. Launch a minimum viable product to get your project off the ground. They firmly believe that now is an amazing time to start a business, and regardless of the outcome, you won't regret investing time into doing something you're passionate about.

Kristen and Anthony also like to share their knowledge with others. Hear audio clips from their profile interview and get other insights and advice on their Plannovation blog (www.cubitplanning.com/blog).

—**JOY ADAMS**

organizations, and also sell their data to customers through a globally accessible platform (WeoGeo 2009). Another example of this kind of small business is Geographic Services, Inc., founded in 2002, which provides an easy-to-use geospatial database providing human geography data drawn from sociocultural research, advanced knowledge of linguistics, and GIS expertise (Geographic Services 2011).

2. Value of Place

Another important perspective in geography is place analysis. Geographers can translate their understanding of place into business practices ranging from creating geography curricula for educators to writing travel guides and developing place-based social media and game applications for smart phones.

Consider the example of Geography Matters, a small business founded by Josh and Cindy Wiggers in 1989 to provide reproducible outline maps, geography textbooks, and map posters to educators and families. Geography Matters has flourished as a home-grown business, and the Wiggers have intentionally kept it as such (Geography Matters 2011). Another example of a small business based on the power of place is Foursquare, a location-based mobile platform that allows people to "check-in" to specific locations via a smart phone, as well as make comments about particular places, bookmark places for future visits, pull up comments by others about nearby venues, and find business promotions near their location. Over eight million people use Foursquare to explore new cities or to inform others visiting their own neighborhoods (Foursquare 2011).

By associating a product with a specific place (e.g., by connecting it with the place of production or drawing on regional identity), businesses can increase the worth of the product. Rusten, Bryson, and Aarflot (2007) argue that place-based products derive a portion of their value from the meanings attributed to a specific place. Marketing strategies often act as a bridge to shift the meanings from the place onto the product. With appropriate training in business principles, geographers can begin to tap the potential of place knowledge for producing their own products and services.

3. "Ecopreneurship" and Interdisciplinary Thinking

In addition to spatial thinking and place analysis, a third geographic perspective focuses on human–environment relationships and working across disciplinary boundaries. As an integrative science, geography is contributing to sustainability in development and business practices. This perspective offers an approach to running a small geography business in a manner that appeals to a growing interest among consumers for sustainable manufacturing, local product sourcing, and environmentally friendly business models. As potential customers begin to shift their purchasing habits toward businesses that operate off of these models, geographers have an opportunity to start businesses that revolve around sustainable, environmentally friendly ideas. If customers begin to select their purchases based on the business model of the company, then the way a geographer runs his or her business may be just as important for its success as the product or service being offered.

In contrast to traditional business models that emphasize maximizing profit for shareholders, Stubbs and Cocklin (2008) suggest that a business can be approached from an entirely different perspective known as the Sustainability Business Model (SBM), which measures success in terms of social, environmental, and economic outcomes. Gibbs (2009) refers to individuals who apply SBM practices as "ecopreneurs" who "combine environmental awareness with their business activities in a drive to shift the basis of economic development towards a more environmentally friendly basis" (55). Gibbs suggests that ecological modernization is becoming increasingly mainstream, which is creating opportunities for sustainable entrepreneurial activity to move into the mainstream as well. Customers' prioritization of sustainable business ventures creates an opportunity for individuals who start sustainable businesses to develop a large customer base.

As these three examples illustrate, geographers can offer valuable services and products based on a spatial approach to economic activity, an understanding of the value of place to people, and a business model that prioritizes environmental sustainability. In each of these areas, geographers can use tools such as GIS to assist in the presentation and delivery of these products and services.

Now that you have some ideas of how geographic concepts and perspectives can turn entrepreneurial dreams into reality, it is time to consider some of the things you can do to prepare for starting your own small geography business.

PREPARING TO START A SMALL GEOGRAPHY BUSINESS

Starting a business is challenging. Not all businesses succeed, and many succeed to varying degrees. However, studies have shown that certain factors can contribute to the success of a start-up company. One of these studies, by Iñaki Peña (2002), looked at the relationship between business start-up success and intellectual capital and found that the two were positively correlated in three areas: human capital (the entrepreneur's level of education, experience, and motivation), organizational capital (structural flexibility for adapting to a market and innovative business strategies), and relational capital (the ability to establish and benefit from business networks). Based on Peña's recommendations, this section covers three activities –acquiring business expertise and skills, thinking creatively about problems and opportunities, and developing professional networks– that aspiring entrepreneurs can do now to improve their chances of success with future start-ups.

1. Acquiring Business Expertise and Skills

Future entrepreneurs can complement their geographic education with courses that will prepare them for running a business. Maguire and Guyer (2004) showed that incorporation of entrepreneurial education into university geography programs was successful in preparing students for the increasingly common "portfolio" careers. Consider taking courses in business/retail geography, economic geography, locational analysis, transportation geography, and population analysis (Estaville et al. 2006). Many of these courses focus on the use of GIS in economic and business decision making.

The following list highlights a few of the most important topics that aspiring small business owners ought to understand:

- Business and employment laws of the state in which the business will operate
- Business plan writing
- Basic marketing
  - Target market
  - SWOT analysis
  - 4 Ps of marketing
- Website development

You can learn about these topics in a university setting, by attending business seminars, or by conducting your own research. Additional details regarding several of these topics will be discussed later in this chapter.

2. Thinking Creatively about Problems and Opportunities

Studies also show that successful entrepreneurs have significant motivation to succeed. One idea for developing a motivated, entrepreneurial mindset, suggested by Carney (2010), is to begin looking for problems for which you could provide a solution and to document at least one of these problems each day. By "problems," Carney means anything that wastes time or energy, causes frustration, or could be handled more effectively. Ashok and Ishu Wadwani's profile offers another example of how a small business offering geographic services stepped up to solve a problem.

This same principle of discovering needs applies when thinking about how your geography abilities might be used to open a cafe distinctive for its regionally sourced meats and vegetables, or a venue with musical programming that pays homage to local culture. Use your geographical imagination as you seek problems to solve or opportunities to enhance standards of living through a small business venture.

3. Developing a Professional Network

Finally, it is important to develop a professional network. The earlier this is done the better, preferably well before the start-up begins. Bönte, Falck, and Heblich (2009) found that the "entrepreneurial process is influenced by the entrepreneur's peers—his or her social network and the corresponding informal contacts" (272). Professional networks are valuable assets for at least three major reasons: a network facilitates access to resources such as capital and labor; it provides information about opportunities and risks, thereby reducing uncertainty; and it gives psychological support, an important driver of productivity (Sanders and Nee 1996).

Professional networks that include other entrepreneurs, investors, and business owners can be developed informally through peers and social contacts, business-oriented social networking sites such as LinkedIn, or online entrepreneur forums and business support centers. Many universities provide small business incubators, programs that provide advice, space, and resources to new companies during or prior to start-up. Some cities host meetings of entrepreneurial groups. University business management departments are a good resource for finding local opportunities.

Having a professional network can assist you in three major stages of the entrepreneurial process: identifying opportunities during the conception stage of a small business; mobilizing resources to exploit those opportunities during the start-up phase; and recruiting specialized personnel if they are deemed necessary (Stuart and Sorenson 2005). A network can also provide your first customer base. Because of its usefulness throughout the entrepreneurial life cycle, it is important to begin to establish a strong professional network in the first stages of starting a small business, or ideally well before the start-up phase. For more on the topic of professional networking, please read Chapter 5 by Tina Cary in this book.

STARTING THE BUSINESS

The trajectory of every business start-up is unique and variable. That said, common start-up flow usually includes determining a problem and a solution (i.e., the product or service the business will provide); developing a team to start the business if it requires more than what you alone

PROFILE 10.2

Ashok and Ishu Wadwani, Owners, Applied Field Data Systems (Houston, Texas)

Ashok and Ishu Wadwani came to the United States in 1970 with two bags and $200. Today, the couple own and operate Applied Field Data Systems (AFDS), a company specializing in field-based Global Positioning System (GPS), GIS, mapping services, consulting, and training.

After obtaining his master's degree in physics from the University of Lucknow in 1963, Ashok landed his first jobs in marketing at the Indian partner offices of U.S. companies such as Perkin Elmer, Hewlett-Packard, and Honeywell. After he and his wife got visas based on their educational backgrounds and obtained green cards, they relocated to Chicago, where Ashok was employed at Central Scientific, a company specializing in lab equipment. While working full-time, both Ashok and Ishu continued to attend school. After Ashok obtained his MBA from the Kellogg School of Management at Northwestern University, he moved among locations and jobs for several years, finally ending up in Houston. By 1984, he had started his own business designing handheld computers for the forestry industry —one of the forerunners of modern GPS.

Because GPS technology was still in its infancy in the mid-1980s, Ashok's entrance into the field came by complete accident. "Early on, I had no clue what GPS even was," he says. Provisions of the Clean Air Act moved him into the realm of fugitive emissions monitoring by 1986, when his company began supplying rugged handheld computers to refineries and petrochemical companies. His clients soon began requesting geospatial information for their emissions data points. "It was a **customer-driven process**," he explains. "GPS technology was developed elsewhere, but AFDS developed the interface not only for petrochemical industries but others." As the company grew, Ishu decided to join the business, giving up her lucrative job in the healthcare industry.

Ashok and Ishu stress that their success did not come easily. While they were able to find jobs quickly upon their arrival to the United States, Ashok notes that the transition can pose a challenge to immigrants not accustomed to American culture. "Asian and European cultures are quite different from American culture, although Americans tend to regard all cultures as similar," he observes.

In twenty-seven years of running their own business, Ashok and Ishu take pride in the fact that they have never had to fire a single employee and have remained debt-free. They strongly believe in **encouraging and mentoring students and new graduates**, and they continue to hire student interns and to offer them flextime so that they can attend classes. Perhaps most importantly, Ashok and Ishu look for people they can trust. Because running a small business means that they both travel often and are frequently away from the office, they must be able to trust employees to get the job done under minimal supervision.

Small businesses operate with fewer financial resources than large companies, and the burden of accountability ultimately rests on the owners' shoulders. However, there is also a great deal of **personal freedom and flexibility**. "You're the boss—you make a commitment, and that's it," Ashok says. For anyone hoping to start his or her own business, he offers some advice: be open to working many hours, be prepared for failures and financial hardships, and be prepared to do odd jobs or **"wear different hats"** within the company. "We firmly believe there are skills we can teach," say Ashok and Ishu, "but we can't teach attitude."

—MARK REVELL

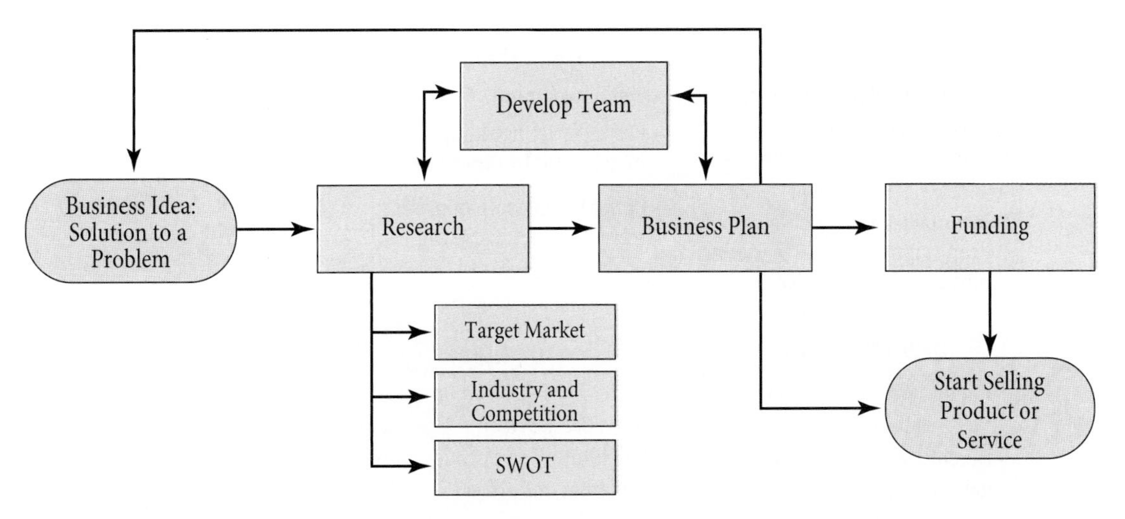

FIGURE 10.1 A common business start-up flow.

can bring to it; conducting research on the target market, needed resources, marketability, and strategies; writing a business plan; seeking funding if needed; and putting the plan into action (Figure 10.1).

The first step is determining what the business will sell. This phase relies on the previously discussed ability to think creatively about problems and opportunities and to develop solutions that meet them. You may find hundreds of problems before you come up with a viable solution to one of them that can be sold for a profit.

Pre-Start-Up Research

Once you have identified a product or service, the next step is to conduct in-depth marketing research and evaluation. I cannot emphasize enough how crucial this activity is to a successful start-up. This research is the foundation on which the business is created, and the information gathered during this stage is the basis for business decisions. First, determine your target market. Are you selling to businesses (e.g., retailers) or directly to consumers? Describe in detail the type of companies or persons who will be purchasing this product, how and why they will use it, and their current purchasing behavior of similar products.

Second, conduct a SWOT (Strengths, Weaknesses, Opportunities, and Threats) analysis. This analysis will help you identify the internal strengths and weaknesses of your company and the external opportunities and threats of the market. Using the SWOT matrix, you can determine where your strengths match up with opportunities, which areas you should capitalize on, and what you need to watch out for. The example SWOT analysis shown in Figure 10.2 reveals that branding the website and company would be a challenge, but one that is crucial to the success of the company's personal solar devices. The challenge with developing the brand stems from the lack of in-person contact with customers. Yet, branding is crucial to the company's success because the personal solar devices it sells are not unique and other companies could sell similar products at a lower price. Therefore, differentiation of the company brand must be made through a unique, attractive, and helpful website and online customer service.

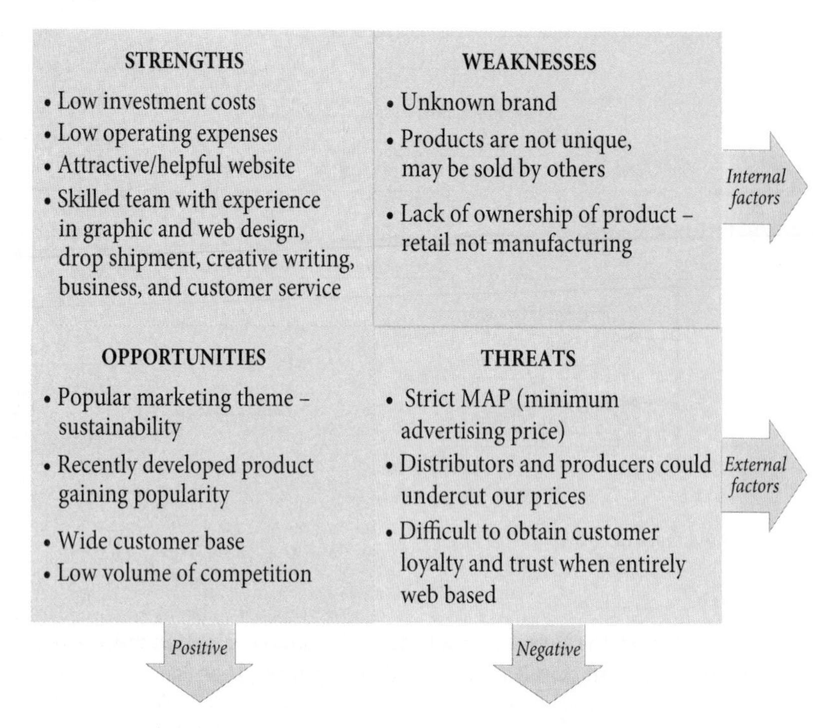

FIGURE 10.2 SWOT analysis for a start-up selling personal solar devices online.

During the SWOT analysis phase, spend time assessing the four Ps of marketing: product, price, placement, and promotion. What makes your product or service unique, how will it be priced, where will it be sold, and how will it be promoted? Consider adjusting your product, price, placement, and promotion in order to compensate for threats or to take advantage of opportunities.

Finally, you need to research the industry and the competition. Determine how large the industry is, how quickly it has been growing, and whether or not it is saturated with products. Ascertain who your primary competition will be and evaluate their sales methods and their hold on the market. By conducting this research prior to starting your business, you can adjust your product or service and your selling strategy before you have invested anything but time. You can also determine if your product is viable and what your competitive advantage might be.

Around the time of conducting this research, you may want to develop a team that will start this business with you. It is possible to start and run a business entirely on your own; however, it may be desirable to find team members who possess skills that complement yours and to bring them on as co-founders. For example, you may want to consider teaming up with someone who has a background in business, or who understands the details of matters such as taxes, legal requirements, payroll, and insurance. Or perhaps you need a product specialist, someone who has worked extensively with the particular product or service you are selling and can troubleshoot or provide customer service for it. In addition to creating a team with varied experience and skills, you will need to ensure that your team can work together, mutually accept the risks associated with a start-up, and approach the business with similar commitment levels. If you have a team, the business plan or another written document should clearly state the expectations for each member's performance, responsibilities, commitment, and ownership.

Writing the Business Plan

The next step is to write the business plan. If you plan to seek funding for your business, a thorough business plan will be required. Even if you do not intend to apply for funding, writing a business plan will help you consider how to make your business successful; it will also prepare you to deal with potential risks as they arise. A study of planning undertaken by small businesses (defined as businesses with less than 500 employees) demonstrated that formal, written planning reduces the probability of firm failure (Perry 2001). Having a business plan is, therefore, a wise investment of time for preparing to start your own business.

A business plan should include sections such as the ones shown in Table 10.1. Every business plan is different, as is the time invested on any specific section. You should incorporate all the main sections into the business plan; however, feel free to adjust subsections and add sections to make the business plan fit your business needs. Plan on spending 5 to 20 hours researching and writing each section.

TABLE 10.1 Common Sections of a Business Plan

- Executive Summary
  - Concise overview of the plan that grabs the reader's interest
  - Tells where the company is and where it is going
- Business Description
  - Explains the basic structure of the business
  - Illustrates how it will be profitable
- Product Description
  - Describes the product or service in detail and the need it meets or opportunity it supports
- Marketing Plan
  - Target market description
  - SWOT analysis
  - Competitive analysis
  - Industry analysis
  - Advertising plan
- Operational and Management Plan
  - Explains the management structure and the responsibilities of invested parties
- Financials
  - Detailed financial projections
    - 1-year projection broken down by month
    - 3-year projection broken down by year
  - Should include current balance sheets and projected income statements

> **SIDEBAR 10.1**
>
> **Resources for Writing a Business Plan**
>
> ***U.S. Small Business Administration***
> Site: www.sba.gov
> What: Provides templates and guidelines for writing a business plan
>
> ***Bplans***
> Site: articles.bplans.com
> What: Collection of articles on every facet of business plan writing
>
> ***New England Business Services***
> Site: www.nebs.com
> What: Step-by-step business plan creation tool

Writing a business plan can be time consuming and generally requires experience and education. Because of this reality, entrepreneurs who have had little training or experience with business plans often hire a professional consultant. The advantages of this option are time savings and the ability to tap into the skills of an experienced writer who can help you think through your business idea more thoroughly than you may have on your own, resulting in a better product and plan. The disadvantage is that you lose the personal connection with your planning. A business plan is very individual—it represents your passion for your product, and some of that is lost when someone else writes your business plan.

If you choose to write your own business plan, extensive help is available online, including templates, detailed instructions, and websites that walk you through the process step by step. Business incubators at many public and private universities can also provide assistance with business plans.

Acquiring Funding and Intellectual Property Rights

Inevitably, you must determine whether or not you need to seek external funding and whether or not to obtain intellectual property rights to protect your ideas. First, consider your funding options. A completed business plan will identify the quantity of capital you will need for your start-up. Can you and your partners fund the company internally? If not, several types of funding are available. Angel investors are individuals who invest their money in small business start-ups in exchange for a piece of equity in the company, often around 15 to 30 percent. They usually have significant experience in the industry or in starting businesses and can bring important insights to a new company.

Venture capitalists are individuals or companies that are looking to invest their money in small businesses for a higher rate of return than they would receive through other investments. In contrast to angel investors, they do not necessarily have any expertise in the start-up's industry. *The Directory of Venture Capital and Private Equity Firms* (Mars 2011) is an excellent source for finding venture capitalists with related interests. A new edition is published annually, and an online database is available to search through the directory for investors who are connected to specific industries, are willing to fund specific investment amounts, or have invested in similar companies in the past. It is important to emphasize that seeking funding through angel investors and venture capitalists often means losing some control over your idea and direction; thus, it is important to consider this option carefully before pursuing it.

Loans from banks and other institutions are also common sources of funding. One of the best ways to apply for these loans is through the U.S. Small Business Administration, which connects funding institutions with deserving small businesses. Finally, many states and private institutions will provide grants to small businesses. Grants are probably the cheapest form of

SIDEBAR 10.2

Resources for Starting a Business

U.S. Small Business Administration
Site: www.sba.gov
What: Podcasts and online courses

Association of Small Business Development Centers
Site: www.asbdc-us.org
What: Find a local small business development center

Internal Revenue Service
Site: www.irs.gov/businesses
What: Tax information, recommended reading, and links to further resources

Gliffy
Site: www.gliffy.com/swot-analysis
What: Assistance and advice on conducting a SWOT analysis

My Own Business
Site: www.myownbusiness.org
What: Free courses in entrepreneurship and a large volume of related resources

FindLaw
Site: smallbusiness.findlaw.com
What: Links to state resources for starting a small business

Directory of Venture Capital and Private Equity Firms
Site: gold.greyhouse.com
What: An online database of financial investors

U.S. Patent and Trademark Office
Site: www.uspto.gov/faq
What: Questions and answers on obtaining a patent or trademark

investment, as the company is often not expected to pay the grant amount back to the giving institution. Applying for grants usually involves showing the granting organization that your business vision aligns with the priorities and values that the grant has been created to support.

Intellectual property rights protect your creative inventions and ideas from being adopted by other companies. Several forms of intellectual property rights can be obtained by applying to the U.S. Copyright Office or the U.S. Patent and Trademark Office. Copyrights give the creator of a literary or artistic work the right to reproduce, display, and sell a creation. They protect works such as novels, music, films, architecture, and compositions. Patents provide the inventor with the exclusive right to produce, use, and sell an invention for a fixed period of time, usually up to twenty years. Trade secrets and trademarks protect product formulas and symbols, respectively (Field 2006). You must determine whether your business is dependent on intellectual property that needs protection; if so you need to take the necessary steps to obtain that protection as soon as possible after starting your business.

CONCLUSION

As described in this chapter, geographic perspectives can solve problems for society or enhance the vibrancy of communities through the vehicle of small business start-ups. The examples discussed in this chapter show how small businesses have benefited from an understanding of the spatial perspective of economic activity, the relationship between place and marketing, and the importance of a sustainable, environmentally responsible business model. Further, geographers have proficiencies, particularly in spatial analysis and geographic information systems, that are in great demand across both the public and private sectors.

Individuals should start preparing for their own business considerably ahead of when they actually intend to start it. During the early stages, focus on expanding your professional network, obtaining relevant education, and looking for problems and opportunities for creatively using your geography abilities. When starting the business, allow time to carry out the planning stages effectively. Although the time required can be lengthy, researching the market, developing a business plan and team, and seeking funding are the foundations of a successful business. Finally, be flexible. Evaluate every positive and negative piece of information gleaned during research, as well as sales dips and spikes once you begin marketing, and adjust as needed. Starting a business is an exciting challenge for geographers and an excellent way to contribute your unique knowledge and skills to others who could greatly benefit from them.

REFERENCES

Bond, D. E. 2005. Geography, business and industry session. *Projecting Geography in the Public Domain in Canada.* University of Western Ontario.

Bönte, W., O. Falck, and S. Heblich. 2009. The impact of regional age structure on entrepreneurship. *Economic Geography* 85 (3): 269–287.

Carney, K. 2010. Starting a GIS business: 2 reasons to be the boss and 4 steps to take now. *Plannovation,* 27 September 2010.

Estaville, L. E., L. Keys-Mathews, B. J. Brown, and W. R. Strong. 2006. Educating business geographers. *Geospatial Solutions:* 28–36.

Field , T. G., Jr. 2006. What is intellectual property? In *Focus on Intellectual Property Rights.* U.S. Department of State, Bureau of International Information Programs.

Foursquare. 2011. About. https://foursquare.com/about (last accessed 15 May 2011).

Geography Matters. 2011. About Us. http://www.geomatters.com/content/about-us (last accessed 15 May 2011).

Geographic Services, Inc. 2011. About Geographic Services, Inc. http://www.geographic-services.net/about/ (last accessed 15 May 2011).

Gibbs, D. 2009. Sustainability entrepreneurs, ecopreneurs and the development of a sustainable economy. *Greener Management International* 55: 63–78.

Kawasaki, G. 2004. *The Art of the Start: The Time-Tested, Battle-Hardened Guide to Anyone Starting Anything.* New York, NY: Penguin Group.

Maguire, S., and C. Guyer. 2004. Preparing geography, earth and environmental science (GEES) students for employment in the enterprise culture. *Journal of Geography in Higher Education* 28 (3): 369–379.

Mars, L. 2011. *The Directory of Venture Capital and Private Equity Firms.* Amenia, NY: Grey House Publishing.

Peña, I. 2002. Intellectual capital and business start-up success. *Journal of Intellectual Capital* 3 (2): 180–198.

Perry, S. C. 2001. The relationship between written business plans and the failure of small businesses in the U.S. *Journal of Small Business Management* 39 (3): 201–208.

Rusten, G., J. R. Bryson, and U. Aarflot. 2007. Places through products and products through places: Industrial design and spatial symbols as sources of competitiveness. *Norwegian Journal of Geography* 61: 133–144.

Sanders, J. M., and V. Nee. 1996. Immigrant self-employment: The family as social capital and the value of human capital. *American Sociological Review* 61: 231–249.

Stuart, T. E., and O. Sorenson. 2005. Social networks and entrepreneurship. In *The Handbook of Entrepreneurship: Disciplinary Perspectives,* ed. S. Alvarez, R. Agarwal, and O. Sorenson, 211–228. Berlin, DE: Springer.

Stubbs, W., and C. Cocklin. 2008. Conceptualizing a "Sustainability Business Model." *Organization and Environment* 21 (2): 103–127.

U.S. Small Business Administration. 2010. http://www.sba.gov/ (last accessed 15 May 2011).

WeoGeo, Inc. 2009. About WeoGeo. http://wiki.weogeo.com/index.php/About_WeoGeo (last accessed 15 May 2011).

11

Going Global: Practicing Geography Internationally

Carrie Mitchell and Mélanie Robertson

You have a desire to see the world and to work internationally. But how can you best prepare for your time abroad, and what can you expect once you get there? While working and living in an international setting can be highly rewarding, it can also be extremely challenging. Being exposed to a new language and different cultural customs, and being away from home can be daunting for many people. At the same time, gaining international work experience can be a way to expand your network, gain valuable and marketable experience, and learn important life skills. In this chapter we will explore the types of work geographers are engaged in abroad, featuring vignettes of geographers working in international careers. We will also discuss the qualities employers are looking for and how students with little to no experience can gain a foothold into an international career. The chapter will conclude with a look at the benefits and drawbacks of working internationally.

But before we begin, let us introduce ourselves. We both hold advanced degrees in geography with a focus on Southeast Asia. Both of us have also worked and studied internationally—Mélanie Robertson with Oxfam in China and the University of Paris in France, and Carrie Mitchell on a project funded by the Canadian International Development Agency in Laos. We both currently work as Senior Program Officers at the International Development Research Centre (IDRC) in Ottawa, Canada. Our primary responsibility as Program Officers is to develop, manage, and supervise research undertaken in the developing world. We believe our mix of academic and applied experience, acquired through various types of work and research opportunities abroad, lends itself well to giving practical advice to young geographers interested in getting into international work. We hope that our discussion and featured experiences will be helpful to current students as well as recent graduates considering an international career.

GEOGRAPHERS WORKING INTERNATIONALLY—A WORLD OF OPPORTUNITIES

Geography as a discipline is experiencing a resurgence in higher education and society as a whole as it is able to tackle issues of local, national, and global significance. As geography promotes interdisciplinary and integrative thinking, it is also highly valued in today's labor market. A representative from a non-profit organization in Colorado put it succinctly: "We're looking for people who are able to think across local and global scales as well as at long temporal resolutions. Geographers are used to being able to think across multiple scales and nest spatial scales together" (Solem et al. 2008, 5).

One of the most important things to remember as you embark on your international career is that few job advertisements use the title "geographer." As a result, geography graduates need to be creative in their search for positions, and they must be able to describe convincingly how their training can assist employers in real-world situations (AAG 2009). The following sections feature opportunities available to geographers in the most common international career paths. This is by no means an exhaustive list of international careers for geographers, but rather a snapshot of the varied employment opportunities available.

Geographers in International Development

Since both of us have established our careers in the field of international development, it makes sense to start with this career path. International development is a very large field, with many types of employment opportunities for geographers. When most people think of international development, they think of the provision of immediate humanitarian aid by bilateral and multilateral agencies such as the U.S. Agency for International Development (USAID) or the United Nations. While this is certainly an essential component of international development, aid agencies also pursue longer-term development projects in a variety of sectors. Geographers may be found working for UN-Habitat, for example, utilizing their geographic skills to analyze human settlement patterns or implementing urban planning projects. Other geographers may find work in the World Bank, either in their Young Professionals Program or as professional staff working on contemporary issues such as climate change. Depending on the organization and the type of employment obtained, staff of bilateral and multilateral funding agencies may be placed permanently overseas, may temporarily travel overseas, or may be based in a head office.

Many geographers also find themselves working for American and international nongovernmental organizations (NGOs), or private foundations engaged in international development work overseas. The website TakingITGlobal (www.tigweb.org) includes a directory of more than 3,000 NGOs worldwide. The kind of placement generally depends on the type of position and the particular organization. The Bill and Melinda Gates Foundation, the Ford Foundation, and the Rockefeller Foundation are all examples of private organizations working in international development. They pursue a wide range of programming, including work directly related to geography such as sustainable development, climate and environment, and urbanization. Finally, for those of you who are more entrepreneurial, or who have a particular in-demand skill, many international organizations are increasingly contracting out work to independent consultants on an as-needed basis (Hindman 2011).

Overall, the field of international development is vast, offering many different types of opportunities for geographers. The profile of Michelle Kooy, a researcher with the Overseas Development Institute in London, offers a glimpse of the profession's rewards and challenges. For a more detailed account of the different types of jobs in the field of international development, take a look at the work of Fechter and Hindman (2011). You may also be interested in visiting websites such as www.aidworkers.net or www.peopleinaid.org.

PROFILE 11.1

Michelle Kooy, Research Fellow, Water Policy Group, Overseas Development Institute (London, UK)

"I always knew that I wanted to work internationally," says Michelle Kooy, Research Fellow at Overseas Development Institute (ODI), Britain's leading independent think tank on international development and humanitarian issues. "I wanted an opportunity to combine my love for research, investigation, analysis, and writing in a real-world setting." As a member of ODI's Water Policy Group, Michelle conducts research for international agencies on aid effectiveness to increase the number of people with access to sanitation and to improve water security.

Michelle received her Ph.D. in geography from the University of British Columbia and her master's in environmental studies from York University, where she developed **research and writing skills** that she says are "very, very important" in her position. In addition to a graduate degree, Michelle's work requires **field-level knowledge of developing regions**. Prior to joining ODI, she had over ten years' experience with research and development projects in a long list of countries in Asia and the Middle East including China, Sri Lanka, Thailand, and Palestine. In her previous job as Director of Urban Programs for Mercy Corps Indonesia, she honed her knowledge of urban water resource management, infrastructure financing, pro-poor service models, and water and poverty issues in developing countries.

Working internationally requires a commitment to a unique lifestyle that has advantages and disadvantages. "You have to realize that you are making life choices when you enter this field," Michelle observes, "because these choices will affect your friends, partners, family, and mobility." Travel is often very sudden rather than planned in advance, and long-term assignments can last up to several months. "One of the hardest aspects is that your partner's life has to be aligned with yours," she adds.

On the other hand, working abroad can be exceptionally rewarding, as it exposes you to new viewpoints, new sources of information, and new perspectives. The process of "**displacing yourself**" through immersion in a country or region is also important for professional development, as it teaches you to cope with being away from familiar settings for an extended period of time and gives you critical insight into the political and physical constraints and conditions of a place. Michelle feels that honing knowledge of the region and/or expertise in a specific policy area or methodological approach is especially important for geographers, whose education is often broadly based. She further observes that **business and management skills** are important aspects of her job that are not often emphasized in academic training, so she advises job seekers to seek opportunities to develop these abilities and to highlight them on their résumés as appropriate.

While there is potential growth on the horizon, the landscape of the international development field is rapidly changing. In the past, relationships could be summed up as "North versus South" or "the G8 and everybody else." But today, "South–South cooperation is becoming more the norm," Michelle explains, pointing out that countries like Brazil, China, and India are now producing their own research and policies internally. Therefore, Asian languages (especially Chinese and Arabic) are in high demand. Because skills such as GIS and modeling, intercultural communication, and field research methods remain important qualifications, geographers continue to be well positioned for employment in the field. Working abroad involves many challenges, including long hours, incessant travel, and difficult life choices. But when you love what you do, it is well worth the sacrifice, says Michelle, adding that the best part of her job is knowing that she is "making a material, physical contribution to people's lives."

—MARK REVELL AND JOY ADAMS

Geographers in the Private Sector

At the (sometimes) opposite end of the spectrum from international development, the private sector offers a wealth of opportunities for geographers wishing to work internationally. The field of geographic information science, for example, offers numerous opportunities, including work within sectors as diverse as distribution and transport, retail, or insurance. Companies looking to expand to new markets overseas need both localized knowledge on the ground and people with specialized skills to plan new retail locations or delivery routes. Many geographers are well suited for this type of work, given their training in locational analysis and geographic information science.

Geographers seeking international work may also find a variety of opportunities in the private environmental sector. Many consulting firms and private companies seek the skills of geographers, posting positions for environmental assessment or environmental monitoring, for example. Geographers work for companies active in environmental remediation or environmental preservation in a host of countries. Geography graduates are also well positioned to contribute to the growing field of climate change adaptation and mitigation, which is receiving increasing recognition, attention, and funding as of late.

As many geographers have the specialized skills in demand by the private sector, you may find work directly with international companies based in North America or with locally based companies in your preferred destination. With the increasing interconnectedness of today's business climate and the emergence of new markets in developing countries around the globe, opportunities for geographers are numerous and expanding. Indeed, as the job market is shrinking in North America, many young graduates, including geographers, are heading overseas to gain valuable business experience (Hudson 2010; Snowdon 2010).

Geographers in the Foreign Service

You will find geographers in the U.S. Foreign Service working in countries around the world in a variety of different capacities (see http://careers.state.gov/officer/career-tracks for more information on the U.S. Foreign Service). Economic geographers may be interested in the Economic Officer stream of the Foreign Service, whereas geographers with a political focus would be well suited for the Political Officer stream. There are many different international career opportunities within the Foreign Service, or with government more generally, that match the varying specializations of geographers. As a student, you can also apply to the student programs of the U.S. State Department (http://careers.state.gov/students), which have positions in Washington, DC and overseas.

We would also like to stress that the Foreign Service, as well as most other international organizations, are seeking employees who not only have appropriate academic qualifications, but also attributes known more commonly as "soft skills." The U.S. State Department has a list of characteristics they look for in employees, known as the "13 Dimensions" (http://careers.state.gov/officer/career-tracks). The list includes qualities such as composure, cultural adaptability, initiative and leadership, resourcefulness, and the ability to work well with others. Although these intangible personal qualities are difficult to quantify, a host of options are available for simultaneously building your soft skills and your international experience. These options will be explained in more detail in the next section of this chapter.

As we have indicated, a wide variety of international work is available to geographers. For more detailed examples of career paths undertaken by geography graduates, we recommend visiting McGill University's Department of Geography website: www.geog.mcgill.ca/other/jobsingeog.html. This page posts short biographies of past students and their current employment, both at home and abroad.

PREPARING FOR INTERNATIONAL WORK

Each organization differs with regard to the expectations of their staff, though there are some common qualifications needed to work internationally. Specifically, international organizations are looking for a combination of education, expertise, experience, and language skills. In the next section, we will delve deeper into these skills as well as where and how to acquire them.

Education and Expertise

With the growing competition for international jobs, graduate degrees are becoming essential prerequisites. It is now commonplace to see organizations looking for candidates with master's or doctoral degrees in a specialized field. Depending on the particular job, international organizations (and domestic organizations with international interests) may also be interested in candidates who have a particular regional focus, for example, Sub-Saharan Africa or Latin America. As one working geographer notes: "Combining regional (Asia) and subject matter (transport) expertise, provided me with excellent credentials to satisfy a growing demand in the North American labor marketplace for knowledge on China and East Asia" (Dr. Daniel Olivier, Economic Analyst, Transport Canada. Personal communication, 18 January 2011). From our experience, combining a specialization within geography (in Mélanie's case it was GIS and in Carrie's case, waste management) and a regional focus is also an excellent way to market yourself to potential employers.

EXPERIENCE Practical experience "in-the-field" is an essential prerequisite for working internationally, as employers are not just looking for degrees but for well-rounded, culturally aware people. International employers, just like their domestic counterparts, seek employees who can get the job done and who do so in an efficient manner. As the work environment in international development is often collaborative, employers generally look for people who can work well in a team, think creatively, and deal well with adverse situations. But how can you prove that you have all these qualities to a potential employer? As those of you who have read job advertisements for international positions may know, the range of required qualifications listed in job ads can be quite exhaustive, even for entry-level positions. However, there are many ways to get into international work that will help you gain the experience you need to compete in the international job market. The profile of Reena Patel, a geographer working for the U.S. State Department, highlights the value of many of the preparation strategies discussed below.

STUDY-ABROAD PROGRAMS Study-abroad programs are an excellent way to gain overseas experience. The challenges of living and studying abroad will demonstrate to employers that you are adaptable, flexible, and open to new challenges—all skills that are vital to succeeding in the workplace. Many universities across North America offer semester-abroad programs at various locations around the world. Information on these programs is generally available from professors, departments, and international centers. Some universities also offer year-long exchanges for students to study at a foreign university while still earning credits toward their degree. In the following account, Carrie details her experience in a study-abroad program:

> *I took part in a semester abroad as an undergraduate. The program allowed a group of 25 students to travel to India, take courses toward their programs of study, and live with Indian families. In addition to taking classes, which ranged from politics, environment, religion, and language, we had the opportunity to travel on class excursions. The expensive airfare was offset by the low housing costs, which resulted in*

PROFILE 11.2

Reena Patel, Foreign Service Officer, U.S. Department of State (Madrid, Spain)

As a first-generation American born to Indian parents who immigrated from East Africa, you might say Reena Patel was born to travel. She got her first taste of working abroad as a Peace Corps volunteer in Ghana. This experience whetted her appetite for a career in international diplomacy, a dream she achieved in 2010 with her appointment as a U.S. Foreign Service Officer in Madrid.

After returning from Ghana, Reena earned a master's degree in technology at Arizona State and a Ph.D. in geography at the University of Texas. While it took her farther from her goal of living abroad in the short term, continuing her education moved her closer to her professional goals and even provided opportunities for travel. Research for her dissertation (*Working the Night Shift: Women in India's Call Center Industry*, published in 2010 by Stanford University Press) took her to Bangalore, Mumbai, and Ahmedabad. While writing, she satisfied her wanderlust by couch-surfing between the homes of friends and family in ten different U.S. cities.

While Reena never aspired to a faculty career, she notes certain similarities between the academic lifestyle and her own. "It feels a little like being a professor," she explains, "but instead of changing classes every semester, you change countries every few years." She also relies on her academic training, the most important aspect of which is, "without a doubt, the **writing skills**." Reena describes her primary responsibility as "delivering demarches": "That's a classy way of calling me the mailman," she jokes. Actually, she is a liaison between national governments, sharing and explaining U.S. policy in exchange for opinions and feedback from the Spanish government. She often drafts detailed reports on issues such as international religious freedom, human rights, and bilateral relations.

When asked about the challenges of her work, Reena is hard-pressed to find examples, but she's quick to note the advantages—at the top of her list is "being out in the world and meeting people I wouldn't have access to otherwise." Her job entails relocating to a new post every two years, and after moving on her own for many years, she especially appreciates that all of the logistics are handled for her. In addition, her salary and benefits provide "a comfortable quality of life."

Reena advises those who aspire to a career like hers to **be persistent**: "You need to be stubborn as well as smart." For example, she applied twice before being selected for a Boren Fellowship through the National Security Education Program and three times before being accepted into the Foreign Service via the Diplomatic Fellows Program (DFP). This program allows Boren Fellows, among others, to submit a curriculum vitae in lieu of taking the dreaded Foreign Service Exam, so Reena believes that it's a great option for applicants who are intelligent but not necessarily the best test-takers. She also encourages applicants to develop a **strong sense of professionalism**, observing that graduate programs typically do not provide sufficient opportunities to develop the **social and networking skills** required for diplomatic work.

Although geographers must learn to effectively market themselves and the discipline, Reena feels that a background in geography provides the proper skills and mindset required for a career in foreign relations. Political and budgetary constraints mean that the competition for positions is very strong, but as long as there's a U.S. government, it will need diplomats: "This is not a job that's going to go away," Reena observes.

—JOY ADAMS

a three-month trip to India costing about the same as one semester at my university. It was the best experience of my undergraduate degree, and it helped me gain a better understanding of a very different country and culture. I now work in India, and my experience of having lived there many years ago has helped me in my current position.

WORK-ABROAD PROGRAMS AND SHORT-TERM OVERSEAS CONTRACTS An excellent way to gain international work experience while completing your degree is to sign up for a student work-abroad program. Information on these programs is available at any university travel center as well as on the Internet. For Americans, we recommend looking into the nonprofit organization InterExchange (www.interexchange.org), while Canadians will probably want to look at the Student Work Abroad Program (www.swap.ca). For a nominal fee, these programs will assist you in getting the visa documents required for working in a selected number of foreign countries on a working holiday. You may end up teaching English, working in a café, or doing some type of seasonal employment. While these may not be your ultimate career goals, the experience you gain from short-term foreign work is invaluable. Moreover, if you apply to take part in these programs right after graduation, you may even be able to obtain short-term work contracts in a field more closely related to your studies, which can be an excellent résumé-building experience.

INTERNSHIPS Internships are some of the best entry-level positions available to recent graduates. In some organizations, interns may be paired with more senior level staff to job shadow them for a period of time. In other instances, interns may work more independently and need to be more entrepreneurial when it comes to defining their positions within the organization. The type of internship you acquire will largely depend on the organization you end up working for.

Although university career centers sometimes post these kinds of positions, you should also look for opportunities through the websites of international organizations such as USAID, the World Bank, the United Nations, or major NGOs. Finally, don't forget professional networks and contacts. Your favorite professor may just know someone who is looking for an intern, and your professor's recommendation might just help you secure the position! This was the case for Carrie, who obtained a one-year internship following the completion of her graduate degree.

I was offered an internship in Laos in one of my professor's projects after earning my master's degree. The salary wasn't exceptionally high, but it was enough to pay off all my student loans and afford a nice little apartment in Vientiane. The work itself was rewarding, but also very frustrating at times. It was the first time I had actually tried to "do" development. In my textbooks it all sounded so straightforward. But here I soon discovered that despite being involved in the same project, my co-workers and I had different ideas about the aims and objectives of the project. This made a seemingly simple project very complex! After my year-long internship, I realized that I still had a lot to learn about development, which prompted me to return to school to continue my studies. I am much more humble now about what it takes to accomplish a development project, which I think is an asset in my current position where I manage research grants.

VOLUNTEERING When we think about gaining employment experience, we tend to think only about paid work or internships. However, volunteering overseas is an excellent way to acquire practical skills and knowledge of a particular locale. If there is a cause you are interested in, or an organization you would like to work with, consider contacting them to discuss volunteer opportunities. Americans will want to consider looking at the Peace Corps (www.peacecorps.gov), a government organization that has been placing Americans in volunteer positions abroad since 1961. CUSO-VSO is one of North America's largest international development nonprofit organizations, and it sends both Canadians and Americans to various countries around the world for volunteer placements (www.cuso-vso.org). Canadian Crossroads International, which a colleague discusses in the following vignette, is a well-established Canadian organization that offers excellent volunteer opportunities for youth and professionals.

> *I did a six-month volunteer placement in Kenya through Canadian Crossroads International (CCI), working with a community-based organization in rural Western Kenya. This placement helped me understand the value of capacity building and the challenges associated with it. I also learned how difficult partnerships and fundraising can be for small local organizations that have minimal staff and administrative capacity. This experience brought the complexity of development home for me and gave me an appreciation for culturally and locally appropriate approaches and technologies. I ended up doing the research for my master's degree in this same area of Kenya, living with the same host family and drawing on a lot of the contacts I had made during my CCI placement. My knowledge of the culture, language, and transportation systems really facilitated my research work and made it an incredibly enjoyable experience for me.* (Heidi Braun, Program Management Officer, IDRC. Personal communication 17 January 2011)

Another colleague we spoke to volunteered in Cambodia while she was doing fieldwork for her master's degree. Rather than working through an established organization, she chose to find volunteer opportunities locally once she was in the country.

> *I did some volunteering while I was away doing fieldwork for my master's degree. Initially, I sought out this kind of opportunity in order to make personal contacts, feel like part of a community, and enhance my social life in a new city. However, my volunteer activities (including craft workshops and English language instruction with low-income youth in their home community) also taught me a lot about my research site and the social dynamics of the people who lived there. Overall, this was a personally and intellectually inspiring experience. This volunteer work resulted in both long-term friendships and contacts for future research projects.* (Dr. Kate Parizeau, Assistant Professor, University of Guelph. Personal communication, 18 January 2011)

These types of hands-on experiences are excellent ways to find out if you really want to work internationally, as they are usually temporary placements abroad with a definitive end date. Depending on the type of program you apply to, you could be volunteering for only a few weeks in the summer or for several years in a particular locale. Of course, longer placements will offer you a better glimpse into the reality of living abroad. Beyond offering a unique and perhaps life-changing experience, the skills gained through volunteer positions can also assist you in your job search later on.

RESEARCH ASSISTANTSHIPS Another avenue for entry into international work is through your professors. If you are a geography student thinking about studying or working internationally, we recommended taking a closer look at what professors in your department (or related departments) are doing. Professors often have research grants, and sometimes hire assistants to carry out some of the work. Sometimes these work opportunities can take the form of desk studies, where you assist the principal researcher by reviewing journal articles and other published reports on a particular topic. In other cases, research assistantships can take place overseas, where you could assist by conducting primary research for your professor's study. In some cases, the research you conduct as part of your assistantship can also be applied toward your degree requirements. Research assistantships may or may not be posted, so if you've performed well in a particular professor's class and are passionate about their research interests, it is worth inquiring with them directly. This type of entrepreneurial attitude is what landed Carrie her research assistantship, as detailed here.

Prior to starting my master's, I had done some informational interviews with professors in the geography department of the school I wanted to attend. I learned that one of the professors had a project in Southeast Asia on a topic I was really interested in. When I started school in September, one of the first things I did was schedule an appointment with the professor to talk about research assistantship opportunities. She told me she had a number of opportunities for students, and let me pick my own topic. Six months later I was in Vietnam, conducting primary research for her project! I was paid for the time I spent abroad, which helped me pay tuition the following year. I was also able to use the research I conducted as part of my thesis. The experience I gained in Vietnam that summer helped me to qualify for an internship in Southeast Asia upon graduation.

Geographers interested in international work can (and should!) combine one or more of these types of opportunities to build their international experience. It is this combination that builds a strong résumé, particularly in the absence of paid work experience overseas. Moreover, in addition to acquiring a combination of experiences, job seekers must be able to successfully present this to prospective employers. Mélanie describes this critical point as follows:

When I applied for my first job after finishing my Ph.D., I didn't have a lot of formal experience in the job market. The particular job I was interested in asked for five to eight years of experience. I then thought about all the time I had spent doing research projects, internships, and volunteering overseas, and when I added that up, it amounted to six years! I applied for the position, and got the job. My "informal" experience was not only critical to getting the job, but also has helped me succeed in the position.

INTERNATIONAL TRAVEL Independent international travel can be one of the best ways to see the world and to learn a thing or two about how it really operates. While not for everyone, and certainly not without risks, backpacking can help one to understand different places and people, think independently, meet new people, and become exposed to new ideas, and, of course, get a sense of whether an international lifestyle is for you or whether you'd feel more comfortable living and working at home.

LANGUAGE SKILLS Language skills are highly valuable to international organizations, and in many organizations bilingualism, or even multilingualism, is a job requirement. Therefore, learning a second or even third language is often essential to an international career. Unless you grew up in a bilingual or multilingual environment, this usually means taking language classes. However, one of the fastest ways to learn a new language is to be in the country where it is spoken. Below some colleagues speak about the importance of learning second languages in their past and current work:

> *My ability to speak French has been a huge asset for my work given that I work with francophone colleagues in West and North Africa. Even knowing some basic Kiswahili impressed my boss in my job interview, as he had done extensive work in East Africa. I learned Luo while in Western Kenya as a volunteer and later while doing research, which has also helped me communicate with colleagues in Africa.* (Heidi Braun, Program Management Officer, IDRC. Personal communication, 17 January 2011)

> *Before conducting my master's research in Argentina, I had taken Spanish throughout my undergraduate degree, but with disappointing results. However, once I had it in my mind that I was going to Argentina for my research, there was nothing that could stop me from reading academic articles and conducting all my interviews, research, and writing in Spanish. Not long into my research, it seemed as if I had spoken the language all my life.* (Suzanne Moccia, Program Manager, Federation of Canadian Municipalities. Personal communication, 19 January 2011)

In a recent (2010) public opinion survey on international education, conducted by NAFSA (Association of International Educators), nearly two-thirds of Americans believed that young people who do not learn foreign languages will be at a competitive disadvantage in their careers (NAFSA, 2010). Learning a language is more than just a means to build your résumé and gain a competitive advantage in the job market; it also provides you with the ability to communicate effectively with people—a skill that will aid you in multiple ways throughout your life.

The Benefits (and Drawbacks) of Working Internationally

The benefits of working overseas are numerous: gaining real-world experiences and cross-cultural learning opportunities, getting a better sense of your country's place in the world, and learning a foreign language. Many of the skills learned by traveling, studying, researching, and working internationally are incredibly valuable to future employers, both at home and abroad. In a study conducted by Hart Research Associates (2010), employers stated that colleges should be placing more emphasis on certain skills and aptitudes to increase graduates' potential to be successful and contributing members to the global economy. Among these skills were written and oral communication, critical thinking and analytical reasoning, and the ability to apply knowledge and skills to real-world settings. Whether you choose a career in North America or beyond, the skills you acquire through international travel, study, work, volunteer, and research opportunities can contribute to your ability to respond to the evolving demands of international and domestic employers. As one working geographer notes, "Living overseas first and foremost expanded my personal and professional network of relationships. By exposing me to adverse situations, it made me more adaptive and responsive to the dynamic needs of today's workplace. It taught me transferable skills that make me a more efficient worker and thinker. Learning about other cultures is also useful in a culturally diverse work environment here in North America and helps me to better relate to some of my colleagues" (Dr. Daniel Olivier, Economic Analyst, Transport Canada. Personal communication, 18 January 2011).

Working internationally may also have drawbacks, something which we feel is important to address as well. People can experience difficult periods of adjustment, particularly when they travel to unfamiliar countries, even if they speak the language. Being away from friends and family is also challenging for many, particularly during extended overseas assignments.

> *Working internationally can be psychologically challenging, particularly when dealing with culture shock, and it can be socially as well as financially draining. It can be difficult to finance overseas work, and spending extended periods of time outside of the country makes it difficult to plan time with family and friends. Language, culture, health, and hygiene standards are all things that can be overwhelming and difficult to surmount for some individuals. Also, living abroad can have a negative impact on personal relationships by imposing a new environment on the partner. The balance between personal and professional goals can be difficult to achieve.* (Dr. Kate Parizeau, Assistant Professor, University of Guelph. Personal communication 18 January 2011)

We should also note that the logistics of working overseas can pose another difficulty. Each country has its own particular visa requirements for different types of visitors (i.e., students and working professionals may need different types of visas, which may have different conditions and periods of validity), and these requirements may change over time. The best way to get the most up-to-date information is to consult the embassy or consulate of the country you wish to travel to. Also, if you are working in a foreign country, you should consider the implications of doing so for your income taxes—both at home and abroad. We also recommend checking in with your country's embassy or consulate when you arrive in your host country, which will keep your name and contact information in case of emergency. The best way to avoid problems overseas is to do your research before you leave. Talk to other people who have traveled to your preferred country or region, read about the country you plan to visit, and examine its rules and regulations. Prior to departing, we also recommend consulting your government travel or foreign affairs department—in the United States, the State Department (www.travel.state.gov), and in Canada, the Foreign Affairs and International Trade Canada (www.voyage.gc.ca)—which offers updated information on the country you are traveling to.

Overall, the decision to travel or not to travel will be one that only you can make for yourself. We hope that this chapter has helped you gain a better understanding of how you can utilize your geography degree in an international setting. We also hope that our practical advice concerning what international employers are looking for in terms of education, experience, and language prerequisites and how you can obtain those skill sets has been helpful. We suggest that you continue exploring your options and talk to people working in your preferred career niche. Informational interviews can be very useful in terms of learning more about your options for employment and about careers you didn't even know existed. You may also want to take advantage of the Ask a Geographer project (www.aag.org) of the Association of American Geographers if you wish to speak with someone in your chosen field. Finally, we encourage you to think about what you're really passionate about and what part of the world you would like to explore, and then to book that ticket and go. Working internationally may well be the best experience of your life!

REFERENCES

Association of American Geographers 2009. Geography: A field of dreams. Available at: http://www.cag-acg.ca/en/index.html (last accessed 27 May 2011).

Fechter, A., H. Hindman, eds. 2011. *Inside the everyday lives of development workers*. Williamsburg, VA: Kumarian Press.

Hart Research Associates. 2010. *Raising the bar: Employers' views on college learning in the wake of the economic downturn*. A Survey among Employers Conducted on Behalf of: Association of American Colleges and Universities, Washington, DC.

Hindman, H. 2011. The hollowing out of Aidland: Subcontracting and the New Development Family in Nepal. In *Inside the Everyday Lives of Development Workers*, ed. A. Fechter and H. Hindman, 169–191. Williamsburg, VA: Kumarian Press.

Hudson, A. D. 2010. What I did when I couldn't find a job. *The Chronicle*. Available at: http://chronicle.com/article/What-I-Did-When-I-Couldnt/66281 (last accessed 27 May 2011).

NAFSA. 2010. Educating students for success in the global economy: A public opinion survey on the importance of international education. Available at: http://www.nafsa.org/publicpolicy/default.aspx?id=23955 (last accessed 27 May 2011).

Snowdon, G. 2010. Graduates look overseas as jobs dry up. *The Guardian*. http://www.guardian.co.uk/money/2010/jul/10/graduates-overseas-jobs (last accessed 27 May 2011).

Solem, M., I. Cheung, and M. B. Schlemper. 2008. Skills in professional geography: An assessment of workforce needs and expectations. *The Professional Geographer* 60 (3): 1–18.

12

Teaching Geography Inside and Outside the Classroom

Susan M. Heffron

Geography is an excellent preparation for a wide range of teaching positions at all levels of education. In recent years, as the field of education has moved toward a stronger focus on accountability and on careful analysis of variables that affect educational outcomes, the teacher has proven time and again to be the most influential school-related force in student achievement (Stronge 2007). Teaching is a challenging and demanding profession, but it can also be a tremendously satisfying experience both professionally and personally.

Employment opportunities for teachers are found almost everywhere, including outside the United States. It is a career that affords you a great degree of mobility in choosing where you would like to live and work. Teaching careers often have some flexibility in work schedules depending on the school or institution. Many teachers enjoy participating in a wide variety of continuing education and professional development opportunities, including travel and advanced degree programs. There is also great potential to develop rewarding relationships with both students and colleagues. There are indeed many intrinsic rewards, but typically the monetary compensation for teaching positions is less than that for other comparable positions.

This chapter focuses on teaching careers in K–12 education, but also examines opportunities in professional settings and informal education that may be less familiar to you. Before taking a closer look at particular career opportunities for geography educators inside and outside the classroom, I would like to share with you some effective teaching practices that could be applicable in any educational setting.

TEACHING GEOGRAPHY AS INQUIRY: ENGAGING STUDENTS IN ACTIVE LEARNING AND INVESTIGATIONS

Geography provides students with a spatial context for analyzing issues and develops modes of spatial thinking that complement the study of topics in many other fields, including math, science, technology, engineering, and social sciences. Geographers ask questions to pursue a greater understanding of the puzzles that result from earth's physical systems, its human systems, and the interactions between them. When geography is taught as a process of inquiry, students learn how to think geographically and apply knowledge and reasoning to a question. Golledge described the nature of geographic knowledge and the changing approaches to teaching it as "a change from inventory-dominated activity to the creation of knowledge generated by emphasizing cognitive demands, such as understanding 'why' and 'how' in addition to 'what' and 'where'" (Golledge, 2002). This means teaching students how to conduct geographic investigations by collecting, organizing, and analyzing data. By engaging in this inquiry process, students will develop skills that can be used throughout their lives.

Inquiry-based teaching has a long tradition in the discipline of geography (Hill 1993; Klein 1995; Solem 2001). All data that has a "place in space," regardless of the context or scale, can be analyzed geographically. For example, water quality data collected from a local river is tied to, and therefore can be mapped to, the collection location. However, students need to consider possible relationships to upstream or downstream collection sites as well as information on land use and population density that may provide additional insights about the data. Students might also question whether the same conditions occur in other rivers in the same state or similar regions in the world.

Teachers who are trained as geographers can easily provide an inquiry approach to explore current events in the classroom. For example, they might ask students to formulate geographic questions and analyze information about the 2011 earthquake and resulting tsunami that devastated parts of Japan. Why is Japan susceptible to earthquakes? How do people adapt to living in an earthquake-prone area? Student investigations on this topic would require looking at maps of tectonic plate boundaries and perhaps maps of the frequency of earthquakes in Japan, and making observations about the safety drills that are conducted in Japan for both earthquakes and tsunamis. They might examine maps of population density to determine how many people live in the most vulnerable areas along the coastlines in Japan or use a geographic information system (GIS) to examine multiple thematic map layers of the region. Students might question how warning systems are used and whether sea walls or other building codes might also be enforced to reduce the negative impacts of earthquakes and tsunamis.

Teaching geography requires organizational and communication skills, in addition to expertise in disciplinary content. Effective teaching also means providing clear and helpful feedback on student work. As a K–12 teacher, you will be required to plan daily lesson activities to meet course outlines and objectives and make assessments to provide feedback on student progress. Geography teachers also often engage students in a wide variety of data collection and analysis activities. Field observations, interviews, and the analysis and interpretation of documents provide sources for informing and improving student research skills.

Geography teachers increasingly need to be proficient in using a wide range of geospatial technology tools (Baker, Palmer, and Kerski 2009). For example, spatial data can now be collected using a variety of digital devices that use Global Positioning System (GPS) technology. These devices range from a GPS-enabled watch to a mobile phone or digital camera with geo-tagging capability, to an actual handheld GPS unit specifically designed for location and mapping tasks. Students can now access many different types of datasets in digital form from websites and online data libraries. These include real-time and archival satellite imagery, which students can analyze to identify land use change over time or spatial relationships that may impact the analysis of a local problem or issue. Esri's Landsat website (www.esri.com/landsat) is an example of a free online collection of satellite imagery that can be used in the classroom by students.

Once data are collected, a wide range of analytical tools can be used that enable students to conduct more sophisticated analyses. Graphs, charts, and maps have expanded to include digital, and sometimes highly interactive, resources generated by huge datasets and remote sensing tools. In addition, digital globes provide users with multiscale viewpoint data permitting students to explore locations "virtually." Likewise, they can analyze and manipulate data using Web-based graphs, maps, and geographic information systems (GIS). Refer to the profile of Paula Ann Trevino, a principal at a charter school in California, for an additional perspective on the learning benefits of GIS for secondary school students.

PROFILE 12.1

Paula Ann Trevino, Grade-Level Principal, Helix Charter High School (La Mesa, California)

GIS is not a part of most high school curricula, but Paula Ann Trevino hopes to change that. She is a strong advocate for bringing technology into the classroom. Her home campus, Helix Charter High School, is evidence of her success. It's one of the very few high schools in America that offers GIS courses.

Although she had originally planned to become a lawyer, Paula Ann had a family and wanted a job that didn't require her to work "70-plus hour weeks." She loved working with kids, so teaching was a natural fit. After receiving her bachelor's degree in social studies from San Diego State University (SDSU) in 1991, Paula Ann was hired to teach at Helix. She eventually moved her way up to Chair of the Social Studies Department, completing a master's degree in education from National University along the way. She earned her administrative credential from SDSU in 2004 and was offered her current position as Grade-Level Principal in 2008. While it is a job that ultimately gives her more leverage in the area of curriculum development, Paula Ann still misses teaching: "Teaching is an art form that allows huge transformations in individuals, one person at a time."

Paula Ann was first introduced to GIS by her nephew, an SDSU graduate student working on a project connecting GIS to teaching. While she was hesitant at first, she eventually "came around" to the idea of introducing the technology into her high school's curriculum after recognizing its enormous **potential for engaging students**. "The possibilities are endless," she says. "I'm constantly amazed at the projects my students come up with." She began researching GIS and seeking out every opportunity she could to **secure grants** for hardware, software, and training. With funding secured, Helix began offering GIS as a component within geography classes, and after a few years, GIS became its own course.

While Paula Ann has been successful at Helix, she notes that on a broader level there are many obstacles facing **the integration of GIS** at the high school level. Chief among them is funding. "For a business, $3,000 might not be much money," she notes, "but it can be significant at a school, particularly where the [GIS class] is just one out of eighty-four classrooms." Another difficulty is finding instructors. Many people who have the technical skills to teach GIS seek higher-paying careers in the private sector, and it is a challenge for full-time teachers to stay abreast of the rapidly changing technology. As Paula Ann points out, subjects like *Huckleberry Finn,* algebra, and Shakespeare are relatively static in comparison.

Paula Ann believes that teachers and administrators need to **take the initiative** if they want to see GIS spread to more K–12 classrooms. Because they are unlikely to be hired specifically to teach the technology, she recommends that prospective instructors first get certified to teach science or social science. Then, "go to a program you know can grow and develop." With the right support in the right places, Paula Ann is optimistic that ten years from now, GIS will be more common in high school classrooms: "We're just scratching the tip of the iceberg."

—MARK REVELL

Engaging students in using data and formulating answers to questions exposes them to the geographical thinking and tasks that geographers employ in their own research and work (Kerski 2003). Knowledge of emerging geospatial tools, online access to cloud-based GIS and satellite imagery, and volunteered geographic data is very important for teachers to use in order to increase what students know and can do in geography. These technologies complement other approaches in the geographer's repertoire, such as field observation, qualitative methods including interviewing and document analysis, and landscape interpretation.

Teaching Geography in Different Educational Settings

Teaching geography today occurs in a wide range of instructional "sites and situations," to coin a geographic phrase. You may find professional teaching positions in informal educational settings such as museums and nature study centers or in more technical settings such as corporate professional development centers where you can teach GIS or remote sensing skills. Geography courses, including human geography, world regional studies, and historical or urban geography, will provide you with the background content knowledge and experience in geographic analysis needed to prepare for a teaching career. You should strongly consider technical courses in GIS and remote sensing to gain experience and skills with those tools. Completing a variety of geography courses (human, physical, technical, and field techniques) will provide greater understanding and knowledge of examples and case studies, all of which will be helpful to you when you enter the classroom as a teacher. Some specific coursework may be required in a program of study for secondary education majors wishing to teach geography. Depending on the organizational structure of the college or university, you may need to consult with the education or teacher preparation program to learn about course requirements in a program of study.

Course selection and preparation may be strongly influenced by degree requirements in colleges of education. The content area and level at which you would teach geography depends heavily on the state in which you plan to teach. In some states, geography is primarily considered a social science, and therefore as part of social studies curricula. In addition, some of the national geography standards are addressed in earth and environmental science curricula as well as in social studies courses. States also vary in whether they assess student learning in geography through end-of-course or statewide assessments. In some states, geography is assessed as a separate subject, while in others it is included in a more broadly defined social studies assessment. These policies are important because assessments often determine the number of courses taught in geography and the emphasis it receives in the curriculum.

As of 2011, geography has a varied position in the U.S. K–12 school curriculum reflecting the different education systems of the fifty states. Each has its own curriculum requirements determined by either state or local policies. The curriculum for geography, social studies, or earth science is typically based on or at least influenced by the publication, *Geography for Life: National Geography Standards* (Geography Education Standards Project 1994). *Geography for Life* identifies content standards that articulate what students should know and be able to do at a range of grade levels. The standards serve as a guide to teachers, curriculum developers at state and local levels, textbook and instructional materials publishers, and assessment developers. In addition, *Geography for Life* has been used as a required text in many preservice geography and instructional methods courses. An updated edition of *Geography for Life* will be available in 2012.

Teaching in K–12 Education

Teaching geography in K–12 education is important and valuable work. Teaching in most K–12 educational settings will require you to obtain a professional teaching certificate in the state where you wish to teach. States certify or license teachers to teach by grade level and subject area. Elementary teachers are different in that they receive an elementary education certificate and are less likely to have any areas of specialization. Middle and high school certification is based on content areas or field endorsements and is issued in accordance with completed coursework. States have different coursework completion requirements for different types of endorsements. Some states issue a specific geography endorsement and require more coursework in geography in order to earn this endorsement. Most states issue a broad social science or social studies field endorsement that licenses teachers to teach all courses in the social sciences. This broad field certification is most often preferred by states, teacher preparation programs, and schools because it provides several advantages, not the least of which is the flexibility for school administrators of hiring one teacher who is legally certified to teach many different courses.

The social science field endorsement typically requires one or two "major" areas of coursework with supporting study in the remaining social sciences. The "core four" social science courses are civics or political science, economics, geography, and history. Based on the taught curriculum in most states, students would be well advised to take the majority of their coursework in geography and history. These are often the most taught courses in middle and high schools, and therefore those for which schools will be hiring teachers.

You may also consider preparing to teach an advanced placement (AP) course in geography. The College Board provides a curriculum and conducts a national end-of-course exam for the AP Human Geography course, which grew from 3,272 exams scored in 2002 to 68,397 in 2010 (Murphy 2007; College Board AP 2010). Specific information about the course is available on the College Board website (www.collegeboard.com).

If you are interested in teaching an AP geography course, be sure to find out if you need any additional preparation to be certified to teach it in a specific state. For example, as of 2004 the state of Illinois requires teachers assigned to AP geography courses to have a "designation" in geography based on 12 credit hours of geography coursework (Illinois State Board of Education 2010). This marked a change in the state compliance requirements, which had previously allowed teachers with the broad social studies field endorsement to teach AP and honors courses. The College Board, state geographic alliances (see more about the alliances in the section on professional organizations below), and some geography departments offer professional development workshops and courses to support new AP human geography teachers when starting out.

K–12 TEACHER PREPARATION PROGRAMS Teacher preparation programs are common in most colleges and universities. Most higher education institutions offer an "accredited" teacher education program that is guided by requirements of the State Department of Education and accreditation organizations such as the National Council for the Accreditation of Teacher Education (NCATE). NCATE is a national accrediting body for schools, colleges, and departments of education authorized by the U.S. Department of Education. NCATE determines which schools, colleges, and departments of education meet rigorous national standards in preparing teachers and other school specialists for the classroom (NCATE 2011).

A student interested in teaching geography at the K–12 level should enroll in the teacher education program and, if available, a geography major degree program. K–12 teacher education programs can take many forms and are evolving as teacher shortages occur in certain states.

At some institutions, the program is completed in parallel with the major degree program, and in other cases, it may be a fifth-year or postbachelor's degree program. Education programs typically involve observations in K–12 classrooms as well as a more extended period of student (or practice) teaching during the course of a semester. An education program usually offers a "teaching methods" course that focuses on how to teach social studies courses, but it may not provide much preparation in teaching strategies specific to geography. For additional support on teaching strategies for geography classrooms, you might seek out a supplemental text such as Phil Gershmehl's *Teaching Geography,* which offers activities and case studies as examples of how to teach geography and spatial thinking concepts (Gersmehl 2008).

CERTIFICATION OPTIONS Teacher certification at the K–12 level is itself a function of geography: each of the fifty U.S. states has its own teacher certification requirements, and there can be a great deal of variance between states. Successfully completing an approved or accredited teacher education program or earning a teaching certificate in one state does not guarantee that you will qualify to be certified in others. There are sometimes reciprocal agreements for certification between states if you successfully complete an accredited teacher education program at a college or university in another state; however, you should always do the research on whether your institution's teacher education program will qualify you for certification somewhere else. In addition to a degree from an accredited institution, some states also require teacher candidates to pass a state content assessment in the discipline that they will be certified to teach. The PRAXIS II exams are an example of subject knowledge-based assessments that are currently required for licensure in approximately thirty-nine states. In addition to discipline preparation requirements, most states also require fingerprinting and possibly criminal background checks as part of their application process. Most state Departments of Education maintain informational Web pages listing all requirements and sometimes the forms needed to complete an application for teacher certification.

Alternative teacher certification programs have become more common at many higher education institutions to assist students who may have already completed a degree and/or are currently working in a field and wish to pursue K–12 teacher certification. First established in the 1980s as a response to projected teacher shortages, alternative education programs are now found in nearly every state and many colleges and universities (Feistritzer 2005). There are currently multiple paths to teacher certification in most states. Meeting with the teacher education faculty advisers is helpful in determining the most efficient and best path to completing the necessary requirements, depending on your current status in a geography degree program.

2 + 2 PROGRAMS Students interested in a K–12 geography teaching degree might also consider completing courses at a local or regional community college that will transfer credits to a four-year college or university. Many community colleges work collaboratively with colleges and universities to facilitate this transfer of course credits and a transition from a two-year to a four-year program. These programs are often referred to as 2 + 2 Programs. You should meet with an academic adviser at the four-year school you wish to attend to make sure the courses you take at the community college will meet the requirements of the program. This path can allow students to take some introductory courses with a lower financial investment while still making progress toward a degree. Undecided students can benefit from taking the time to explore introductory geography and GIS courses as well as some introductory education courses to determine whether they wish to work toward either of these bachelor-level degrees.

Teaching at the University Level

If you have completed or are near completion of your undergraduate degree in geography, you may also consider applying to a graduate program. Pursuit of a master's or doctoral degree often involves teaching undergraduate courses as part of a graduate student's assistantship in a department. Graduate students may not have had any real teaching preparation prior to their teaching assignments. This can be challenging for many, and while most departments now offer mentoring, support, and some preparation to beginning graduate students, often the graduate student is learning how to teach "on the go." University-wide teaching centers can offer valuable support to new college instructors. In addition, the AAG offers two books supporting college-level geography teaching strategies, *Aspiring Academics* (Solem, Foote, and Monk 2009) and *Teaching College Geography* (Solem and Foote 2009). Refer to these books for many insights into preparing for a teaching career in higher education.

Teaching in Professional Development Settings

Some corporate, government, and research-based institutions hire trainers to teach new software applications and techniques to their professional and research staff members. Teaching in this environment can be both challenging and very rewarding. Participants are often paid to attend, are highly motivated, and are very interested in how the material presented can be put to immediate use in their field of work. GIS applications or online mapping tools, as well as geographic approaches, analysis, or field techniques, are topics that may be taught in these courses.

The participants in work-related professional development or training settings often want the new approaches or tools to be directly tied to their work objectives and tasks. Your preparation may involve learning what types of projects, tasks, or work the participants do in order to make direct connections as to how the geographic approaches and tools will help them with their daily work assignments. Geography students with a technical GIS or online mapping background, as well as those who are experienced in new field techniques, may seek out opportunities to teach professional development workshops for their current employer. Organizing and effectively presenting professional development workshops requires detailed planning and should incorporate many of the same active teaching strategies that are appropriate for adult learners.

Teaching Geography Online

Online computer-based instruction has changed dramatically in scope and interactive capabilities since its inception. Today a wide range of courses and programs are available online to students and to instructors who may be interested in teaching in virtual venues. These opportunities may include adjunct instruction of courses at higher education institutions—those required for either a degree or a technical certificate program. Some GIS certificate programs are taught almost entirely online. A great deal of variety still exists in the design of courses and technological capacities in online and virtual instruction.

Much like face-to-face teaching, online teaching is characterized by advantages and disadvantages for both instructors and students. If you are considering teaching online, learning the basics about course management systems such as Blackboard or Moodle would be good preparation prior to teaching a course. Online instruction also requires time management skills in order to keep the students and the course on track. One online instructor interviewed for this chapter provided the following insights:

> The opportunity to interact with participants from diverse locations is especially rewarding. As an [online] instructor, asynchronous (and to a lesser extent synchronous) online learning environments allows me to give more considered feedback and responses to participant inquiries. This ability was especially helpful in my early teaching endeavors. Having the opportunity to fact-check and enhance my replies allowed me to be a more confident, and hopefully, more effective instructor. (D. Ward, personal communication 4 February 2011)

The beginning elements of effective online instruction are grounded in the basics of effective traditional instruction. The seven principles for good practice in undergraduate education identified by Chickering and Gamson (1987) provide documented components of best practices in traditional classrooms. Student–faculty contact, cooperation among students, active learning strategies, prompt feedback, high expectations, time on task, and respect for diverse learners and learning styles all contribute to effective instruction. More recently, Chickering and Ehrmann (1996) reviewed these seven effective instructional strategies in light of the current tools that can be used to achieve effective instruction using technology. These basic building blocks for using technology tools in effective instruction should be included in any online professional development designs. Studies of online distance learning in geography can also guide you toward effective practices (DiBiase and Rademacher 2005; DiBiase and Kidwai 2010).

Informal Educational Settings

Opportunities to teach exciting geography content is not confined to formal or online educational settings. Many different venues host informal education experiences, courses, workshops, and field studies. Put on your creative thinking hat and recall places you have toured, visited, or taken an informal education class or workshop. Museums, zoos, aquariums, botanic gardens, environmental centers, state and national parks, travel and tourism companies, community-based recreation centers, and even some retail specialty stores have established education programs and offer different types of geography-related courses or workshops. Many museums, zoos, aquariums, botanic gardens, and environmental centers maintain education programs or departments and pursue educational grants and awards to assist them in offering courses and workshops. Organizations such as these may offer project-based teaching opportunities supported by grant funding. Others, such as the National Center for Atmospheric Research in Boulder, Colorado, maintain an education and outreach program that offers face-to-face as well as online educational experiences for both physical and virtual visitors to the Center, in addition to more formal classes for school field trips. In these settings, coursework in physical geography, human–environment interactions, and technical courses in GIS and remote sensing would provide a good basic preparation.

State and national parks also host informal educational programs and can be an opportunity for getting started in a teaching career. Park rangers and interpreters conduct lectures and field walks or hikes on a regular basis. Descriptions of permanent and seasonal positions are provided on the National Park Service website (www.nps.gov/personnel). REI, a retail outdoor equipment company, offers workshops in geocaching, GPS skills, reading topographic maps, and orienteering at their local stores as part of their REI Outdoor School program (www.rei.com/learn). Even cruise lines are now offering educational programs that may provide employment opportunities for a teacher with geography expertise. For example, the National Geographic Society has partnered with Lindblad Expeditions to offer educational expeditions to many parts of the world (www.expeditions.com), and positions with different human and physical geography specialties on staff are available. Sometimes these teaching opportunities are offered more sporadically and viewed as nontraditional instruction, but they can offer an excellent forum for testing your interest (or passion) in the preparation and teaching skills required for a future teaching career without committing to a full-time teacher preparation program.

Geographers who enjoy sharing their expertise with others can often find teaching opportunities in informal educational settings for both human and physical geography topics and techniques. Searches on organizational websites for jobs, careers, and educational opportunities will help you learn more about possible teaching opportunities in these informal education settings. Professional networking through association listservs, peers, and social networking are all good strategies for finding teaching and education related positions. Geographers are teaching when they lead urban field walks, provide lectures for community groups and organizations, and serve as mentors on GIS projects being completed by classrooms or youth programs. You may also consider volunteer experiences as a way to introduce yourself as a potential instructor in informal educational settings. In her profile, Tara Gettig, an educator working for a state park in Pennsylvania, further explains the value of professional training in geography and how experiential learning opportunities can enhance the qualifications of aspiring environmental educators.

PROFESSIONAL ORGANIZATIONS SUPPORTING GEOGRAPHY EDUCATION

The National Council for Geographic Education (NCGE) focuses on supporting and improving the teaching and learning of geography. It is an excellent professional association for geography teachers at all levels. The NCGE publishes an academic journal, the *Journal of Geography*, as well as a journal focused on K–12 teaching methods entitled *The Geography Teacher*. NCGE sponsors the National Conference on Geographic Education each year where workshops, paper sessions, and field studies provide excellent professional development and networking experiences for teachers. More information about membership and activities is available online at www.ncge.org.

The Association of American Geographers (AAG) is a professional association of geographers located in Washington, D.C. The AAG sponsors many faculty development and teacher professional development projects as well as the creation and dissemination of teaching and learning materials for both instructors and students. The AAG publishes two academic journals: *Annals of the Association of American Geographers* and *The Professional Geographer*. The association also has a geography education specialty group consisting of members with a special interest in all aspects of geography education. More information on the AAG is available online at www.aag.org.

The National Geographic Society's Education and Children's Programs division is closely tied to the National Geographic Education Foundation, which funds a state-based network of geographic alliances. Most states have a functioning geographic alliance that will provide resources, mentoring, teacher professional development workshops, teaching and learning materials, and often additional course and field study opportunities for K–12 teachers in the state. A list of contacts for the geographic alliances is available on the National Geographic Society's education website (www.nationalgeographic.com/education).

PROFILE 12.2

Tara Gettig, Environmental Education Specialist, Pine Grove Furnace State Park (Gardners, Pennsylvania)

An avid outdoor recreationalist, Tara Gettig has always loved public lands. But it wasn't until she spent a summer as an intern at Shenandoah National Park that she discovered her passion for education in an outdoor setting. "I enjoyed watching visitors have an 'A-ha' moment," she recalls. Realizing that she could have a career that combined her interest in the natural world with her love of maps and curiosity about places, she joined the staff of Pennsylvania State Parks after completing her master's degree in geography at Shippensburg University in 2007. Tara worked in the agency's central office in Harrisburg for several years, where one of her duties was to help develop visitor maps. In May 2011, she became the environmental education specialist for Pine Grove Furnace State Park.

As the park's sole educator, Tara must be **knowledgeable about a wide variety of subjects**. "I use my geography background to connect our park visitors and general public with Pennsylvania's natural resources," she explains. As a geographer, she also strives to instill in visitors a "deeper sense that geography is beyond being just about places on a map to what makes those places special." Pine Grove Furnace State Park is celebrated as the midpoint of the Appalachian Trail, so Tara wants to promote awareness of the trail's historical and recreational importance as well as an appreciation for the nation's public lands and the challenges of creating and preserving them. "So many people live in close proximity to the trail, but they aren't aware that it's there and that there are hundreds of people traveling it," she observes.

Tara relies heavily on her undergraduate training in communications, as both **public speaking and writing** play large roles in her work. Her outdoor experience also comes in handy, as the agency sees recreation as a means of fostering care and concern for natural resources. She's currently developing plans for an ecotourism program that incorporates hiking, biking, and paddling with an exploration of regional land uses and industries to demonstrate how geographic characteristics fit together to create a unique sense of place.

Reflecting on her professional development, Tara feels that her internships were critical in helping her to decide what kinds of work she likes and dislikes in addition to helping her build her professional network. She encourages geographers to **seek out work-based learning opportunities** to help them jump-start their careers, citing her agency's "excellent" internship program as one example.

Although Tara had no formal academic training in education, she believes the most important qualification for being an outdoor educator is "a **passion for teaching** others, for reaching out to others and getting them excited about what you're excited about." If you have that, she says that you can learn the necessary skills and information. Pennsylvania State Parks also offers its educators Certified Interpretive Guide training through the National Association for Interpretation [NAI], as well as many other training opportunities. To learn more about NAI's training and certification programs, visit http://www.interpnet.com.

Public lands have always faced funding challenges, and traditionally, educational programs have been among the first to be cut in tough economic climates. Fortunately, many forward-thinking leaders are becoming increasingly supportive of environmental education, recognizing that it is a key element in building support and creating stewardship for public lands. While many entry-level positions are short-term or seasonal, Tara notes that there are full-time opportunities for those who do well and are persistent. Employees who are willing and able to move around will find that the abundance of temporary positions allows opportunities to make a living traveling among the country's most beautiful places. Despite the current hiring constraints within all levels of government, Tara believes that we are building the next generation of workers to support public lands: "We see our interns as the future of the state parks."

—JOY ADAMS

SUMMARY

For those with a love of learning, teaching geography offers not only an opportunity to help others gain knowledge and understanding, but also continuing learning opportunities for the teacher. Teachers have incredible potential to guide student learning and help model and teach critical thinking skills in students.

Today, opportunities for teachers are found in more places than you might imagine, both within and beyond traditional academic and formal educational settings. Teaching careers outside of K–12 and higher education may be found in museums and nature centers, as well as corporate professional development programs and virtual instructional settings. New highly interactive websites have the ability to meet just-in-time learning needs and have expanded the range of instructional opportunities for teachers. As opportunities to teach and learn expand, so too do the venues where you might consider teaching geography and sharing this engaging and dynamic discipline with others.

REFERENCES

Baker, T., A. Palmer, and J. Kerski. 2009. A national survey to examine teacher professional development and implementation of desktop GIS. *Journal of Geography* 108 (4–5): 174–185.

Chickering, A. W., and S. C. Ehrmann. 1996. Implementing the seven principles: Technology as a lever. *AAHE Bulletin* 49 (2): 3–6. Available online: http://www.tltgroup.org/programs/seven.html

Chickering, A. W., and Z. F. Gamson. 1987. Seven principles for good practice in undergraduate education. *AAHE Bulletin* 39 (7): 3–7.

The College Board. 2010. 2010 AP human geography student grade distributions. http://apcentral.collegeboard.com/apc/public/repository/2010_HumanGeo_Score_Dist.pdf. Last accessed 30 August 2011.

DiBiase, D., and K. Kidwai. 2010. Wasted on the young? Comparing the performance and attitudes of post-adolescent undergraduates and postbaccalaureate students in identical online classes. *Journal of Geography in Higher Education* 34 (3): 299–326.

DiBiase, D., and H. Rademacher. 2005. Scaling up: How increasing enrollments affect faculty and students in an asynchronous online course in geographic information science. *Journal of Geography in Higher Education* 29 (1): 141–160.

Feistritzer, E. 2005. *Profile of alternative route teachers.* Washington, DC: National Center for Education Information.

Geography Education Standards Project. 1994. *Geography for life: National geography standards.* Washington, DC: National Geographic Society Committee on Research and Exploration.

Gersmehl, P. 2008. *Teaching geography*. New York, NY: Guilford Press.

Golledge, R. G. 2002. The nature of geographic knowledge. *Annals of the Association of American Geographers* 92 (1): 1–14.

Hill, A. D. 1993. Geographic inquiry into global issues. *OAH Magazine of History* 7 (3): 73–76.

Illinois State Board of Education. 2010. Guide to requirements for certification, endorsement and assignment of teachers, school service personnel, and administrators. Illinois State Board of Education's (ISBEs) Certification Website at http://www.isbe.net/certification. Section VIII: 27. Last accessed 4 April 2011.

Kerski, J. J. 2003. The implementation and effectiveness of geographic information systems technology and methods in secondary education. *Journal of Geography* 102 (3): 128–137.

Klein, P. 1995. Using inquiry to enhance the learning and appreciation of geography. *Journal of Geography* 94 (2): 358–367.

Murphy, A. 2007. Geography's place in higher education in the United States. *Journal of Geography in Higher Education* 31 (1): 121–141.

National Council for the Accreditation of Teacher Education (NCATE). 2011. Website at http://www.ncate.org. Last accessed 30 June 2011.

Solem, M. N. 2001. A scoring guide for assessing issues-based geographic inquiry on the world wide web. *Journal of Geography* 100 (2): 87–94.

Solem, M., and K. Foote, eds. 2009. *Teaching college geography: A practical guide for graduate students and early career faculty.* Upper Saddle River, NJ: Prentice Hall.

Solem, M., K. Foote, and J. Monk, eds. 2009. *Aspiring academics: A resource book for graduate students and early career faculty.* Upper Saddle River, NJ: Prentice Hall.

Stronge, J. H. 2007. *Qualities of effective teachers, 2nd Edition.* Alexandria, VA: Association of Supervision and Curriculum Development.

13

Geography Careers in Consulting

Susanne C. Moser and Angela J. Donelson

OUR JOURNEY INTO CONSULTING: AN INTRODUCTION

"Oh really?" was the question that rang in my transition from academia into consulting. I (Susi) didn't know it at the time; I had made no plans. Contrary to all the "how-to" book advice (which I read only much later), I didn't carefully consider the transition, nor what it would mean to run my own business. But here I was: One late afternoon in December 2007 I was catching up on phone calls while driving home from a briefing with Boston's mayor, Thomas Menino, about what climate change might mean for his city. While in the meeting, a long-time colleague from a research program run by the California Energy Commission had left a message to ask for my advice on some social science research priorities the state should include in its strategic research plan. So, I called him back, gave my advice, and as we chatted, he told me that the state was considering hiring a social scientist (on a long-term contractual basis) to help them strengthen the social science side of their climate change portfolio. "Oh really?"

By the end of that conversation, a serious offer to be that social scientist was on the table, and I was considering it equally seriously. For months, I had felt the growing urge to change jobs; at the time I was on research leave from the National Center of Atmospheric Research (NCAR) in Boulder, Colorado, one of the country's premier research laboratories for atmospheric science. Moreover, my partner had just declared a desire to go back to graduate school to obtain a doctorate—the only school that offered the desired degree was . . . in California.

What followed was a month-long frenzy to gather as much information as I could about the logistics of consulting. I dove blindly into the cold waters of contract negotiations, found out I had to be a subcontractor to a big company to make it happen, had no money to pay a lawyer to explain the legalese I was asked to sign, lost sleep over deciding on a company name, and gave my notice to NCAR. By the end of February 2008, we moved.

In retrospect, and several years wiser and farther down the road of consulting, I marvel at the way the stars aligned and how courageous and willing—or really, naïve but lucky—I was in letting go of a secure income with lovely benefits and vacation, and yet, how the need for a change, more fulfilling work, and the desire to be my own boss catapulted me out of that "safe" place, through all the fears and not knowing into a new career.

For me (Angela), the move into consulting was similarly serendipitous, but I arrived there from a different place and for different reasons. It was actually my second time venturing into this world—but this time I was venturing out as a self-employed consultant, in contrast to my first consulting role as a community planner employed with a consulting firm. Toward the end of my dissertation work, and while teaching a course on regional development at the University of Arizona, a nonprofit organization asked me if I would help them with an affordable housing project. I had already worked with them in my prior role at the U.S. Department of Housing and Urban Development for five years, advising them on related issues, so without thinking too deeply about it, I accepted the offer. Soon they asked if I could help with additional projects and their program evaluation. I agreed—the projects were interesting and varied, my work schedule was flexible, and I felt like I had made an important contribution to their antipoverty work. So after about a year of trying it out, I thought, "Why not try to grow my consulting business and these opportunities further?"

That reflected a departure in my thinking. My experience with a consulting firm a decade earlier was unlike that of some other graduate school colleagues I knew. Many had really enjoyed working in that culture, travelling to new places to work with a diversity of clients, often on large-scale projects. My experience felt different. I didn't have much of a hand in what I perceived to be the "fun" stuff—the conceptual design and interaction with the clients. To me, the work felt too "boilerplate" in nature. Yet, that job was valuable in giving me experience in the field and confidence that I could actually do this kind of work.

Consulting was a better fit this second time, not only because I had more creative license as the sole proprietor of my firm, but also because of the flexible nature of the work, which aligned well with my personal goals. I had decided not to interview for academic jobs in geography. I knew starting out in the academic field would require a demanding full time schedule that would be difficult on my family. I didn't want to move—my husband had a job he loved. I had two-and-a-half year old twins, who, born very prematurely, still needed a lot of my time. In addition, I had already invested a dozen years in my career field. I didn't want to start over.

As I continued solo, other contracts soon followed via word of mouth, with foundations, state and local government, and small and large community development nonprofit organizations. In my past five years on this path, I have really enjoyed the diversity of the work, and especially the ability to work with interdisciplinary teams to learn new things. Those projects with a strong research-based or evaluation component have particularly appealed to the social scientist researcher in me. Perhaps most importantly, this work has fulfilled my desire to be of service, to support and strengthen the work of organizations working in the community and economic development sector, and to act in a space that helps bridge theory and practice.

These are our stories of how we, the authors of this chapter, arrived in the world of consulting: unplanned, by happenstance, and through some strange window of opportunity when needs for change and greater flexibility were met by an income-generating prospect to apply our skills, experience, and talents. Many consultants, like the two other human geographers profiled in this chapter, get there in just such unforeseen ways. Others make it a deliberate plan, either directly out of graduate school or in midlife. Some keep one leg in whatever profession they are already in (e.g., academia, a nongovernmental organization) and add on consulting projects when opportunities arise. Some stay in consulting for the duration, while others move back out into other employment situations.

In this chapter, we will give our perspective on the world of consulting and embed it in related research and experiential literature. Interestingly, not much of that literature exists in geographic journals, maybe because physical geographers—who seem to be far more at home in the consulting world—work there but don't write much about their experience (for a notable exception, see McLellan 1995), and because human geographers are still a relatively rare occurrence there. In other disciplines, too, research interest in the work life of the consultant is a relatively recent occurrence (since the 1990s; see Jacobson, Butterill, and Goering 2005). In the third section we focus on the skills and perspectives that make geographers valuable and "marketable" consultants. We give examples of the kinds of projects in which human geographers can get involved. As the fourth section will show, more than geographic skills are required to be a successful consultant as a geographer. This mix of skills makes this form of earning a living both challenging and interesting, one that offers constant opportunities for professional growth and learning. We sum up these challenges and benefits in the final section.

A PEEK INTO THE WORLD OF CONSULTING

Before we ever worked as consultants, we didn't really think about what it would be like, at least not to be one on our own. And then we sort of "fell into it" and found out. But many readers of this volume may be interested in, or are already considering, this career option and may want to know.

Our experiences reflect those of sole proprietors; that is, of owners and managers of very small businesses (typically with no further employees). Consulting can also be done in the context of employment with a larger firm. That route tends to pose less risk to the individual (in terms of benefits, health insurance, steady work, greater stability, and often more opportunities for entry-level geographers) than the self-employment option we have chosen. In comparison, self-employment offers distinct opportunities for becoming an independent, well-rounded professional exercising individual creativity, work-life balance, and a greater span of control, and for creating one's own work teams. We talk more about the specific opportunities and challenges associated with "solo" consulting in the fourth section and the conclusion of this chapter.

First, we will address some of the questions often asked by individuals who are interested in consulting. We feel it is important to consider what it actually is like and how it feels to work in the consulting world, not just to think about the topics and issues that characterize the work of consultants. We examine specific examples of using geography in consulting in the third section.

Who Wants, Needs, and Uses Consultants?

Another way of asking that same question is: Who are our clients? Or: Who would pay for our services? These are important questions to consider, as trends in those segments of the economy and of society will tell you something about the market you enter. Over the past two decades, consulting as an industry has grown tremendously, despite ups and downs in the economy. This growth can be related to larger economic and government trends—including downsizing, outsourcing, increased job mobility, specialization, the general growth of the "knowledge worker" segment of the economy, devolution of government, and so on (Flynn 2000; Kitay and Wright 2007; Donnelly 2009; Hicks, Nair, and Wilderom 2009).

Thus, consultants are hired by just about everyone—governments at all levels (international, federal, state, regional, and local), from multinationals to smaller businesses, from not-for-profit organizations (e.g., environmental advocacy groups, social service organizations, educational institutions) to private foundations. Research-oriented consultants can also obtain research funding from government agencies and collaborate with academics.

Why Hire a Consultant?

Consultants provide services that one hires for a certain purpose, for a limited time. It is often not a long-term commitment, but some consulting firms have long-term relationships (read: contracts) with certain clients. For example, big consulting corporations may have five-year renewable contracts with federal or state agencies to provide certain services—research projects, ongoing meeting support, accounting and bookkeeping, and mapping services to name just a few examples. Small consulting firms may build such good rapport that the client may come back time and again for a particular kind of expertise. Yet other projects are more short-lived, "freelancing" opportunities doing assorted creative tasks (e.g., as a writer, editor, web designer, analyst, or cartographer) without necessarily having a commitment to an ongoing engagement with the same client.

Why can't these jobs be done in-house? To the extent a particular field is changing rapidly, firms and agencies may need very specialized knowledge that is simply not available or coming in fast enough with the normal turnover of staff. Temporary workers are often more cost-efficient; thus where there are budgetary limitations, a consultant is a better option than either hiring or retraining staff. Consultants coming in from the outside can also help bring about change inside by increasing legitimacy for a planned change or helping make tough choices (Jacobson, Butterill, and Goering 2005).

What Roles Do Consultants Play for Their Clients?

This brings us to the question of what roles we can play for clients, or what expectations clients may have of consultants. Reading across the disciplines and fields, we found a plethora of descriptors, and between the two of us, we have probably been called upon to play any one of these: We have served as interpreters or translators of science; we have provided an extra hand to get things done; we have served as coaches, teachers, trainers, expert advisers, fact finders, and technicians; we have been questioners, change agents, and stimulants of change; we have been bridge builders, storytellers, and advocates; we have been—or have been perceived as—partners, allies, and mentors; we have held up mirrors, probed, and prodded; we have been leaders, scouts, and steady hands; and while we haven't cured any illnesses (real or metaphorical), some of our clients probably wished we would be "witch doctors" or bring some other kind of "magic" to intractable problems. (All metaphors have been written about in Fritchie 1988; van Es 1993; Fischer 2001; Czarniawska and Mazza 2003; Jacobson, Butterill, and Goering 2005; Fincham et al. 2008; Sturdy 2009.)

This wide range of roles speaks to the diversity of tasks and demands, to the challenges and rewards, and to the power dynamics that can exist between consultants and clients. In the past, probably a greater power differential existed between consultant and client, whereby the consultant was assumed to serve the client without question, to be at his or her beck and call. Nowadays, there is a greater tendency for a more symmetric and dynamic relationship between the two (Czarniawska and Mazza 2003), with clients and consultants being engaged in a more interactive model of knowledge sharing, co-production, and learning (Jacobson et al. 2005; Fincham et al. 2008), and a mutual sharing of responsibility (Poulfelt 1997). I (Susi), for example, have experienced power sharing and a lack thereof in one and the same project; at times helping to shape the direction of the project and at others being excluded from key decisions about its direction, and instead resigning myself to what I considered less of a co-leadership role, or even just "damage control." How these things work out in the end often depends on the quality and strength of the relationship between the individuals involved, and the political sensitivity of the project in question.

What Is It Like to Be a Consultant?

So what is it like or how does it feel being a consultant? We raise the question because we probably weren't prepared for some aspects of that experience. Certainly coming from academia, I (Susi) wasn't prepared for the sudden shifts in how some other academics viewed me—my previous affiliation with NCAR had given me a prestige that did not always carry over to my being a consultant, even though what I actually worked on was virtually identical. While I previously did a lot of work with the media, I fell off the rolodexes of most journalists with whom I had ongoing relationships, probably because of some shift in perceived "independence." At the same time, many of my long-term collegial relationships remained just as strong as ever. They continued, didn't change in quality, and some led to exciting new projects. At the same time, I wasn't entirely sure how to behave in the new world I entered.

For me (Angela), I found that relationships remained strong with those I'd previously served in a capacity-building role at the U.S. Department of Housing and Urban Development, such as foundations and local nonprofits. Not surprisingly, my role with employees at the federal agency with whom I had worked shifted. Although I remain in close contact with a few, I think many have not been quite sure how to frame relationships with former employees who went into consulting, like I did.

The literature on consulting is quite explicit on these matters. Identity issues, existing in a liminal space and being "other," are common challenges (Prus 1996; Czarniawska and Mazza 2003; Collinson 2006; Fincham et al. 2008). As an expert consultant, you are not an academic, but also not a practitioner (policymaker, decisionmaker in government, executive in a firm, etc.); you are not an insider of the client organization but yet are required to know enough to serve it well; you don't exist in anyone's organizational culture, and you constantly cross boundaries. By not "belonging" in this way, some consultants experience themselves as being invisible, and some feel disrespected because of it. Others feel their disciplinary background or training becoming blurred, or disappearing altogether. Living outside the social structures of academia or government also results in some consultants feeling that they have little status, power, or control, at the same time that this lack of defined hierarchical structures, expectations, and obligations liberates them to unfold as they wish. Thus there is quite a demand on the consultant to engage in what Prus (1996) calls "identity work"—to resist stigmatization, stereotypes, and categorization through reinventing and redefining oneself and through establishing and maintaining one's credibility and reputation; to counteract invisibility through one's professional presences; to actively rework one's self-image; and to strengthen one's self-esteem through internal and external support (Czarniawska and Mazza 2003).

What Are Some of the Challenges that Consultants Face?

Not so much because of the particular tasks, but because of the liminal space in which they find themselves, consultants often operate in situations of ambiguity, sensitivity, and bounded rationality (limited knowledge). They may be technically superior to the client in some ways, but are not the only ones having relevant knowledge; they definitely don't have much power over decisions, and they can be fired at a moment's notice. Despite all good intentions and contractual mechanisms that aim to form a "mutual" relationship, this ideal does not always materialize (Poulfelt 1997). The particular status of the consultant as an expert outsider (from the client's perspective), as a businessperson with responsibilities to herself (or a company) on the one hand and to the client on the other, and as an independent—often solo—worker, makes consulting a high-wire act with few safety nets underneath.

Consulting involves juggling many different demands and making trade-offs (e.g., times of intense work load versus family obligations or commitments to one's personal growth). Ardichvili (2000) describes several of the common dilemmas consultants face, and we can attest to them from our own experience. Consulting offers more flexibility to use your talents, skills, ideas, interests, and time, and a greater sense of control over your work and destiny, but it also demands greater responsibility. Whatever turns out great and whatever fails, it is all *yours*. Success requires a lot of self-discipline, self-direction, and self-motivation. Consulting can also entail great excitement over the kinds of projects and work one does, but also anxiety over whether or not there will be a steady flow of income, new projects, and cash flow. This implies a need to budget time and money well because much of what happens in consulting is not steady. Consultants thus face considerable uncertainty, which takes some getting used to; developing a modicum of mental and emotional stability; and having reasonable contingency plans and support from others.

Starting out as a new consultant also raises the dilemma of how to market and establish yourself with your unique qualifications by going it alone versus partnering. Closely related is the dilemma of focus versus diversification: How does one balance the need to be seen as an expert, gaining credibility, and continuing professional growth with the need to stand on a broader footing for income stability? We and other consultants have found that networking with just about anybody and partnering with others who have complementary skills are among the most effective strategies to navigate these tricky questions. We certainly agree with Ardichvili (2000) that "independent consulting, as opposed to highly specialized [. . .] jobs, provides an opportunity for becoming a more rounded individual. This is achieved not only through structured learning, but also through on-the-job acquisition of new skills, trying out new approaches and methods" (p.138). But who you want to become, and what you want to learn and do, need continual revisiting as markets and fields of inquiry change, your experience grows, and client needs shift.

Poulfelt (1997) and Ardichvili (2000) also describe a range of ethical dilemmas that affect all consultants: How much professional effort is too little or too much in light of the need for profitability? Where is the right balance between what might constitute an optimal approach versus the client's budget? How much knowledge, information, and detail are enough to meet the client's needs? How should one balance one's own interest versus the client's interest? Who actually is the client (for example, in consulting for government, is it the agency that pays for the contract, the person who is your point of contact, her boss, or the wider public served by the government agency?), and thus whose needs is the consultant trying to meet? What is the reputation of the client, and thus how may that reputation impact your own? What does the project you are being invited to contribute to, what are its motives and underlying agenda, and what is its potential use? Finally, how close can and should one get to the client in terms of having potentially confidential insider knowledge and in building trustful, even personal, relationships versus being impeded by that knowledge and proximity in relationships? Awareness of these issues maybe the first and most important answer we can offer as advice here, as early alerts can prevent many more challenging problems later.

We also suggest doing careful research into the project and client if you don't already know them well *before* you commit to a piece of work. If things go wrong or something emerges over time that doesn't seem right, it is good to seek advice from friends, former advisers, and colleagues to whom you can speak in confidence. You may also look for insights from reliable online sources or professionals (such as legal aids, or conflict and mediation experts). If you first worked for a bigger company, a government agency, or a university, you may have had the opportunity to attend professional development classes that taught you valuable skills and background (e.g., on various laws, antidiscrimination statutes, effective communication in conflict situations). Such courses are also sometimes offered by the Small Business Administration, the local Chamber of Commerce, or a local community college (see also Francis Harvey's Chapter 15 on professional ethics in this volume). Clearly, there is no simple resolution to ethics questions in consulting by relying on some professional code of conduct or by turning to a higher authority. Instead, resolving these ethical dilemmas is contingent on your values, your role, your contract, and the specifics of the context to which you must answer. In the end, you are the ultimate authority who gets to make the judgment call and has to decide. The truth is, it is both your burden *and* your freedom to meet that challenge in a way that allows you to look in the mirror and be at peace.

What Does a Consultant Do on a Daily Basis?

As for the mundane, basic, logistical aspects of consulting, we can only say two things: First, consultants, especially in small businesses like ours, do virtually anything and everything—from copying to bookkeeping, from contract negotiation to original research, from applying for small-business loans to submitting research grants, from designing their own website to fixing the bugs on their computers, from making their own travel arrangements to giving high-powered speeches, from writing business plans to drafting brochures, and from submitting project reports to publishing research papers in high-impact journals. Logistical help on how to get started and how to run a business is ample online and can be found in Table 13.1. In larger firms, just as in government or academia, work is often more specialized for any one individual consultant because there are experts focusing on different subject areas and also administrative, web, and other support personnel.

TABLE 13.1 Some Useful "How-To" Resources on Consulting

Books

Block, P. 1999. Flawless consulting: A guide to getting your expertise used. San Francisco, CA: Jossey-Bass.

Considered by some the "consultant's bible," this guide takes you step by step through the stages of consultant–client interaction, and pays particular attention to the changing roles consultants play nowadays: not just providing advice and expertise, but building the client's capacity.

Cohen, W. A. 2001. *How to make it big as a consultant*. New York, NY: AMACOM.

From the back cover: A consultant acts as marketer, salesperson, subject expert, legal adviser, accountant, negotiator, and more. Understand what clients think they need (and analyze if it's what they really need), market your business, work through the basic elements of any project (proposals, pricing, contracts, scheduling), solve clients' problems, structure and run the business (all the legal, tax, and insurance issues), and grow and evolve to become an outstanding consultant.

Cohen, W. A. 2006. *The entrepreneur and small business problem solver*. Hoboken, NJ: John Wiley & Sons.

The go-to resource for budding entrepreneurs and small-business owners, this revised and updated edition covers everything from getting a start-up loan to insuring your business, from recordkeeping to doing marketing research, from introducing a new product to protecting your business against rip-off.

Marsh, P. D. V. 2001. *Contract negotiation handbook*. Surrey, UK: Gower Publishing.

(*continued*)

Books *(continued)*

For those who have never done contracting before and those who need to know the finer details, this book is a useful handbook to have near your desk. The successful planning, execution, and conclusion of contract negotiations relieve heartburn and contribute to financial sanity, legal protection, and profitability. If nothing else, it's a primer on bargaining and negotiation.

McLellan, A. G. 1995. *The consultant geographer: Private practice and geography*. Waterloo, ON: University of Waterloo.

This book addresses the challenges and opportunities for those interested in consulting careers in geography. Based on three decades of experience in combining an academic career and private consulting in geomorphology, McLellan argues that academic geographers and the academic curriculum have not been giving enough attention to the opportunities for geographers in the private sector. He calls for a new breed of geographer, which he characterizes as a 3D Geographer fully involved in design, development, and decision making.

Weiss, A. 2009. *Getting started in consulting*. 3rd. ed. Hoboken, NJ: John Wiley & Sons.

Another very helpful step-by-step guide to getting started. Which to pick from these? Thumb through them and see what reads best and answers most of your questions.

Articles

Ardichvili, A. 2000. Critical dilemmas for the independent consultant. *Consulting Psychology Journal: Practice and Research* 52 (2): 133–141.

The author discusses four dilemmas—and strategies to navigate them—that appear to anchor the experience of many independent consultants (regardless of their field): flexibility and control versus responsibility; enjoyment of the work versus anxiety; marketing versus partnering; and focus versus diversification.

Druckman, D. 2000a. Frameworks, techniques, and theory. *American Behavioral Scientist* 43 (10): 1635–1666.

A good overview of the challenges and opportunities for social scientists in the consulting world. Not specific to geography, but largely reflective of our experience.

Jacobson, N., D. Butterill, and P. Goering. 2005. Consulting as a strategy for knowledge transfer. *Milbank Quarterly* 83 (2): 299–321.

Authors differ on what roles they favor or emphasize that consultants play. This article discusses various models of how research gets utilized in practice, and promotes an interactive, collaborative model of the interactions between consultants, academics, and decision-makers through the various stages of the consulting process.

Littlefield, C. N., and E. Gonzalez-Clements. 2008. Creating your own consulting business. *NAPA Bulletin* 29: 152–165.

A great little article on how to start and how to operate a consulting business by two anthropologists. The disciplinary issues may be different, but the great suggestions for consulting apply regardless.

Stumpf, S. A., and R. A. Longman. 2000. The ultimate consultant: Building long-term, exceptional value client relationships. *Career Development International* 5 (3): 124–134.

It's surprising how much one can learn from these short, general articles. If nothing else: your challenges and concerns are neither new nor yours alone.

Troper, J., and P. D. Lopez. 2009. Empowering novice consultants: New ideas and structured approaches for consulting projects. *Consulting Psychology Journal: Practice and Research* 61 (4): 335–352.

Not specific to psychology at all, this article suggests how faculty with some consulting experience and ongoing practice can prepare their students for the world of consulting.

Websites

Small Business Administration
http://www.sba.gov/category/navigation-structure/starting-managing-business
25 Steps for Starting a Consulting Business
http://www.teaching.nmsu.edu/resources/bookstore/kd25steps.pdf
Just for example.... *"geo-economic consulting" (from a university course on consulting, with lots of links to other geographic consulting firms)*
http://faculty.washington.edu/krumme/resources/consult.html

Second, if you're a sole proprietor like we are, we suggest thinking carefully about whether you *should* be doing it all yourself. Your specific geographic expertise and all the years of training that went into it may be better spent on the substantive aspects of your work, and a little help from outside can free you up to apply yourself most efficiently. Unless you have friends who can offer their help for free, outside assistance costs money, and that needs to be carefully balanced against your income and resources. However, sometimes it pays to go for "the best" even if you have to pay dearly for it.

GEOGRAPHIC CONSULTING: A WORLD OF SKILLS AND PERSPECTIVES

Let us turn now to the substantive question of what geographic consultants do. Geographers can offer consulting services in an extremely diverse range of employment areas—for example, in qualitative and quantitative research, in program design and evaluation, in facilitation and strategic planning, and in training. While consulting as a career choice has been long established in physical geography (such as in soil sciences, water resources management, and environmental compliance) (see the excellent book by McLellan 1995 for more insights on physical geography consulting), human geographers are increasingly carving out unique specializations. Because the latter is less well known, we highlight below our own experiences and those of Kate Edwards of Englobe (Englobe.com) and Victor Konrad of NorthWingConsulting (NorthWingConsulting.com).

The work of consultants trained in geography is very diverse. Victor Konrad, a former tenured academic professor in geography and administrator of the Canada-U.S. Fulbright Program, has built on his decades of expertise in research, teaching, academic administration, and cultural resource management to deliver consulting services in regulatory compliance, strategic planning, facilitation, fund-raising, and international program development to nonprofit organizations and government agencies. Kate Edwards, an experienced cartographer and geographer who developed a geography division within Microsoft, is a geopolitical and cultural strategist. She consults with private sector companies such as Google, Microsoft, and computer game developers, helping them to adapt digital content to the critical geopolitical and cultural expectations of local end users.

I (Angela) am a former regional and community planner and federal nonprofit trainer/capacity building specialist who worked with a wide variety of housing and economic development organizations in the U.S.-Mexico border region. I have used this experience, along with my Ph.D. training as an economic geographer, to deliver consulting services to small and national nonprofits, foundations, and state and local governments that need help planning, implementing, and evaluating community and economic development strategies and programs.

And I (Susi) am a geographer working primarily on topics and with approaches that fall within the human-environment tradition of geography (www.susannemoser.com). Trained first in physical geography and then for a Ph.D. in human geography, I now work mostly on the human dimensions of global change. My research and consulting business with governments at any level, nonprofits, and academic and philanthropic institutions focuses on three pillars: human vulnerabilities and responses to climate change, communication, and improved science-policy interactions and decision support.

The Value of Geography in Consulting for the Corporate Sector, Nongovernmental Organizations (NGOs), and Government

Geographic thinking in consulting has great value for both the private and public sectors. Geographers have unique perspectives, skills, and training that enable them to bring forth environmental, geopolitical, and/or sociocultural contexts to the center of projects. This understanding of how to analyze and integrate these diverse aspects makes geographer-consultants essential methodological and theoretical "anchors" in interdisciplinary teams.

Kate Edwards, for example, found that in working for Microsoft, her master's level geographic training prepared her to develop her own geopolitical strategy department, while coordinating content with software developers. Her division became the central resource for coordinating digital content strategies for mapping software, such as Encarta, and other products, such as digital games. By the end of her tenure at Microsoft, every product leaving the company had a geopolitical line item, ensuring it had addressed cultural sensitivity needs. She found that usage of items in digital content such as colors, names of geographic locations, and gestures were critical to products that were culturally sensitive and responsive, and didn't have to be remade each time a product was released in a new language. Her current consulting practice addresses this very important need—for companies to avoid making expensive political and cultural mistakes.

Victor Konrad similarly views his training as a geographer as instrumental in integrating and interpreting the work of interdisciplinary teams. For example, in his consulting work on the *Historical Atlas of Canada*—a project that spanned work from the late 1980s through the 1990s—he brought the interpretation of historical and prehistorical resources to the public, through pulling together the work of a team of historians and anthropologists. Similarly, his training as a geographer has helped him effectively assess bi-national cultural issues in working with government agencies on Canada/U.S. border projects. He believes geographers have an essential role in applied work on border issues because they are capable of bringing together the work of scholars in related disciplines, such as in political science and law. More generally, the multidisciplinary training of geographers, which uniquely looks at the intertwined relationship of numerous physical and cultural dimensions of issues, enables geographers to link and synthesize the various disciplines engaged in addressing problems or issues better than experts from other disciplines.

I (Angela) have also found my training as a geographer to be essential in integrating cross-disciplinary and cross-cultural work. Through training in graduate school, my understanding of how to define and analyze regional impacts has helped me structure bi-national (U.S.-Mexico) economic and community development projects and community development program evaluations. I also have found geographic analysis skills essential in demonstrating to clients how various spatial scales (such as the individual, neighborhood, community, U.S. regional, and bi-national regional levels) are critical in one's work, and how these scales often overlap and intersect, creating different ranges of impacts. Similarly, my training in geographic information systems (GIS) has given me tools that allow clients to see the impact of their work in ways they had not originally conceived. For example, as a consultant to the State of Arizona's Early Childhood Development and Health Board, I worked with an interdisciplinary team of educational policy and community psychology consultants to map physical and socioeconomic assets. We implemented use of zip-code level stress data and GIS mapping of community assets, which are informing how the agency plans, prioritizes, and distributes funding for early childhood programs. Although the state regional boards had not thought about using geographic mapping in their scope of work, they have found it to be extremely useful in identifying impediments and resources to childcare and early childhood development programs.

Some of my work (Susi) involves bringing the insights of human geography to governments and NGOs in the context of planning for and managing the impacts of climate change. For example, I have found that very few planning, public health, and natural resource management agencies know how, or do not have the capacity, to conduct social vulnerability assessments (using Census data, quantitative methods, and GIS, or even just qualitative data), nor do they know how to use them in their priority setting, planning, and tracking of impacts over time. In one project, I partnered with two other NGOs and a postdoctoral student from a local university and conducted such vulnerability assessments for two local counties in California. We used this background research to inform workshops with local stakeholders and to initiate local planning discussions on how to prepare for and deal with the now unavoidable consequences of climate change. One common experience in this kind of work is that people who don't know much about what climate change might mean for them, and who are quite skeptical of this global problem at the start, come to see how focusing on reducing vulnerability can yield benefits to them right here and now, creating the community they really want, and prepare them for the future.

The Skills Geographers Bring to Consulting

As geographers, we developed many of our basic research and analysis skills in graduate school. The process of writing a thesis or dissertation has been especially helpful in consulting: They taught us how to envision and contextualize a problem, how to write and communicate effectively, how to digest and translate large amounts of information, how to research information to address a specific problem, how to use tools for analysis (such as GIS or statistical analysis software), and how to interpret results in meaningful ways.

Yet, all of us have "layered" these basic skill sets onto the technical expertise we have carved out in a particular content area. None of us moved straight from graduate school into consulting as a first career. As independent consultants, we each first developed an area of interest and expertise, and gained practical experience in that field for at least a decade. We also have found it takes years of work experience to develop some of the "soft" skills required in consulting. Victor notes that the skills of management, planning, fund-raising, and communicating effectively with others have taken life experience to develop. I (Angela) believe that networking is another essential "soft" skill learned through work experience before consulting. In particular, "unlearning" the jargon-heavy academic writing style many of us worked so hard to acquire in graduate school is essential in working with nonspecialist clients or with others bringing different expertise. There is an art and earned skill in writing clearly, succinctly, and yet substantively for different audiences—and in consulting, you get paid for doing that well! Finally, each of us has acquired additional skills (e.g., public speaking, running effective meetings, managing the business affairs of our operations) that have little directly to do with geography, but are essential to successfully running our own businesses.

Kate occasionally receives calls from geography students wanting to work in her field as consultants. She advises them to step back, to take time to find out why they like geography and what they are passionate about, and to begin working with a company first—perhaps as an analyst, a strategist, or a marketing person. The key, she said, is to learn first how businesses think and to develop one's niche by infusing their own unique thinking as a geographer into it. She advises students who are venturing into the work world to find geopolitical and geocultural openings and weaknesses in a company's strategy, so as to improve the company's work. Regardless of whether one works for business, government, or NGOs, we believe it is important to understand the industry and gain work experience within it before consulting for it (see also Chapters 6 and 7 on working for government, Chapter 8 on the corporate sector, and Chapter 10 on small businesses other than consulting firms). I (Susi) happened to start my business at the beginning of a major economic recession. Without some 15 years of prior experience in my field, an extensive network of colleagues, and ample connections outside of academia, I doubt I would have been able to launch a thriving consultancy.

Examples of Settings Where Geography Has Opportunities for Growth in Consulting

As interdisciplinary social scientists whose work is grounded in place and space, geographers often possess an ideal skill set for consulting. However, Victor doesn't see many geographers preeminent in the social services end of consulting. He says that geographers aren't usually the first to come to mind as consultants on specific issues, such as poverty issues—those would be sociologists—nor do they feature big on cultural issues—those would be anthropologists. Yet, the opportunities for geographers interested in consulting are there. In community and economic development, for example, geographers are ideally suited to help organizations think about, analyze, and evaluate their work at various scales of impact, and how these scales and impacts intersect. In the fields that integrate multidisciplinary teams, the skill sets of geographers, such as in GIS proficiency, make them indispensable to the success of the consulting team.

Social science consulting, including human and human-environment geography in the context of climate change, global change, and sustainability, is nothing short of a "growth industry." I (Susi) say this with more trepidation and concern than with mere joy and satisfaction over my business prospects. The human dimensions of climate change have long been neglected in academia and by funding agencies, and they still have no foothold in government agencies. This means that much societal response to climate change will go forward without the insights from geography and related disciplines—unless and until more of us work on the ground, informing policy and decisionmaking as consultants and otherwise (Moser 2010). Thus, the need for applied, place-specific research and useful information is almost endless, and geographer-consultants can find needs to fill around every corner.

LIVING AS STRANGE HYBRIDS: POSSIBILITIES FOR COLLABORATION AND CAREER-MAKING

In the preceding section, it became clear how much our work involves collaboration with other consultants, agencies, a wide range of clients, and researchers in geography and other disciplines. It is one of the great benefits and ongoing learning opportunities that keep consulting fresh and interesting. As self-employed, independent consultants, we have developed significant technical competence and enjoy a great deal of autonomy; our projects are varied, calling on a range of skills, talents, and creativity; we feel we are of service to something we believe in; and our work at times has direct and tangible impact and influence on real people, in real places.

These benefits notwithstanding, we all have worked in arenas other than consulting before (experiencing the pros and cons there), and we may want to move elsewhere some time down the road. That work may take you out of your niche into a new area of expertise, or out of the particular format and venue of your work (consulting) into another (e.g., government or academia). When that time and those opportunities will arise, you never know, but if you have even an inkling of a hunch that this is something you may be interested in at some point, it's good to get a foot in the door, keep doors open, or start laying the foundation for a new career path.

I (Susi), for example, have always lived as somewhat of a "strange hybrid" between academia and the world of practice, and continue to do so—whether or not I ever return there. I recently acquired a half-time position at Stanford, enjoying academic research and the interaction with colleagues, while maintaining my business and continuing applied work with governments and NGOs around the country and world. To keep that academic door open, I publish in peer-reviewed journals (often using the experience and insights gained in my consulting work), apply for research grants, attend and speak at scientific conferences, review manuscripts for journals, and serve on National Research Council committees and scientific advisory boards. Opportunities for such hybrid work lives exist in many areas. They are immediately useful to keep things new, and they may offer options in the future you barely know you are planting the seeds for now.

CHALLENGES AND BENEFITS FOR THE SELF-EMPLOYED CONSULTANT: A CONCLUSION

Consulting on one's own is not always an easy path. For those who are self-employed, developing one's business takes time and effort, and barriers can easily discourage entry (Braddock and Sawyer 1985; Stumpf and Longman 2000; Littlefield and Gonzalez-Clements 2008; Weiss 2009). As suggested earlier, you may consider working in a larger consulting firm or a different sector first to gain some experience, grow your professional networks, and solidify your skills before venturing into solo consulting. Consultants who have written about the challenges faced by consultants, especially those on their own, note that job insecurity or significant variation in income can be a primary concern, as one often has too much work or not enough (Druckman 2000b; Viola and McMahon 2010). This can threaten work-life balance, be difficult to manage financially, and counteract the desire for a schedule that accommodates other part-time work or family needs, which is why some consider sole-proprietor consulting in the first place.

One can easily be drained by having to learn and attend to the "business" aspects of consulting—such as billing, collecting delinquent payments, developing proposals, learning how to build in adequate overhead costs on projects, filing taxes, and setting up the right business structure. Some view the need to attend to the business of consulting as a threat to the intellectual challenges posed by consulting (Druckman 2000b). Others might enjoy the diversity of tasks and challenges, becoming a "more rounded individual."

Professional isolation—especially for small consulting businesses like ours—is another challenge, but one can counteract this "occupational hazard" by participating in professional networks, working collaboratively, developing professional relationships with other consultants, and maintaining networks with clients and academic colleagues. Such networking opens new learning opportunities, provides exposure and a chance to market oneself, and sometimes leads to new contracts.

Consultants who have straddled the academic and consulting worlds can sometimes feel sidelined to marginal professional status by traditional academics. Others in social science consulting have echoed this concern (Druckman 2000b; Fincham et al. 2008). From our perspective, academics often seem ambivalent about, or even hostile to, consulting in geography. Why is this so? Staeheli and Mitchell (2005, 359) find that researchers in academia often see applied geography as being "atheoretical or ad hoc rather than systematic . . . it is said, therefore, to run the risk of complicity in projects that undermine its own normative values." Humanist and radical scholars in geography may see consulting as a lesser way to address social problems and ineffective for understanding the "big questions" about important global phenomena, such as economic restructuring and environmental change (Staeheli and Mitchell 2005). Yet, as Staehli and Mitchell make clear, what is "relevant" work in geography is intensely subjective and always context specific. In our experience, consulting is an opportunity to influence social policy directly and immediately, which we would not have easy access to in a traditional academic career. Moreover, one's methods and approaches need not be atheoretical at all. Rather, as others have noted (e.g., Druckman 2000a), we find practice-relevant research done in the context of consulting to be a true test of relevant theory, and thus an opportunity to advance the development of grounded theory.

On the other side of these challenges stand numerous benefits, and many would argue they outweigh the drawbacks. If satisfaction with one's work is fundamentally a function of you (1) knowing yourself, (2) understanding the role that consultants can play in specific contexts, and (3) knowing how to make or improve the fit between you and the role of the consultant, then there is a great chance that you can reap the benefits of this career choice (Webb 1993). Rewards include the opportunity to develop practical applications of theory and methods, and the chance to see directly how your good research is being put to practical use for making a valuable difference. We especially enjoy helping solve social problems for clients with whom we like working.

As Webb (1993) notes, consulting includes the freedom and creativity to operate free from organizational norms; to have significant autonomy and variety in tasks, clients, people, and projects; to be of service; and to demonstrate technical competence. Similarly, Viola and McMahon (2010) observe that consulting is a good fit for individuals possessing and seeking to further personal characteristics such as authenticity (which they observe to be one of the most important characteristics for building trust with clients), a desire for continual learning, and a tolerance for ambiguity, empathy, flexibility, and self-confidence. As a career choice, Druckman (2000a) notes that research consulting provides the opportunity to become your own "think tank." It affords those interested in social science research with the opportunity to advance a field of study and to operate between the boundaries of theory and practice (Czarniawska and Mazza 2003; Jacobson, Butterill, and Goering 2005; Fincham et al. 2008). So, if you're not afraid of actually having impact and influence, consulting could be just the ticket.

REFERENCES

Ardichvili, A. 2000. Critical dilemmas for the independent consultant. *Consulting Psychology Journal: Practice and Research* 52 (2): 133–141.

Block, P. 1999. *Flawless consulting: A guide to getting your expertise used.* San Francisco, CA: Jossey-Bass.

Braddock, B., and D. Sawyer. 1985. Becoming an independent consultant: Essentials to consider. *Nursing Economic$* 3 (6): 332–335.

Cohen, W. A. 2001. *How to make it big as a consultant.* New York, NY: AMACOM.

Cohen, W. A. 2006. *The entrepreneur & small business problem solver*. Hoboken, NJ: John Wiley & Sons.

Collinson, J. A. 2006. Just "non-academics"? Research administrators and contested occupational identity. *Work, Employment & Society* 20 (2): 267–288.

Czarniawska, B., and C. Mazza. 2003. Consulting as a liminal space. *Human Relations* 56 (3): 267–290.

Donnelly, R. 2009. The knowledge economy and the restructuring of employment: The case of consultants. *Work, Employment & Society* 23 (2): 323–341.

Druckman, D. 2000a. Frameworks, techniques, and theory. *American Behavioral Scientist* 43 (10): 1635–1666.

Druckman, D. 2000b. The social scientist as consultant. *American Behavioral Scientist* 43 (10): 1565–1577.

Fincham, R., T. Clack, K. Handley, and A. Sturdy. 2008. Configuring expert knowledge: The consultant as sector specialist. *Journal of Organizational Behavior* 29: 1145–1160.

Fischer, D. 2001. Value-added consulting: Teaching clients how to fish. *Curator* 44 (1): 83–96.

Flynn, P. 2000. The changing structure of the social science research industry and some implications for practice. *American Behavioral Scientist* 43 (10): 1578–1601.

Fritchie, R. 1988. So you want to be a consultant? Or consultancy by any other name! *Management Learning* 19 (2): 105–108.

Hicks, J., P. Nair, and C. P. M. Wilderom. 2009. What if we shifted the basis of consulting from knowledge to knowing? *Management Learning* 40 (3): 289–310.

Jacobson, N., D. Butterill, and P. Goering. 2005. Consulting as a strategy for knowledge transfer. *Milbank Quarterly* 83 (2): 299–321.

Kitay, J., and C. Wright. 2007. From prophets to profits: The occupational rhetoric of management consultants. *Human Relations* 60 (11): 1613–1640.

Littlefield, C. N., and E. Gonzalez-Clements. 2008. Creating your own consulting business. *NAPA Bulletin* 29: 152–165.

Marsh, P. D. V. 2001. *Contract negotiation handbook*. Surrey, UK: Gower Publishing.

McLellan, A. G. 1995. *The consultant geographer: Private practice and geography*. Waterloo, ON: University of Waterloo.

Moser, S. C. 2010. Now more than ever: The need for more societally relevant research on vulnerability and adaptation to climate change. *Applied Geography* 30 (4): 464–474.

Poulfelt, F. 1997. Ethics for management consultants. *Business Ethics: A European Review* 6 (2): 65–71.

Prus, R. 1996. *Symbolic interaction and ethnographic research*. Albany, NY: State University of New York.

Staeheli, L. A., and D. Mitchell. 2005. The complex politics of relevance in geography. *Annals of the Association of American Geographers* 95 (2): 357–372.

Stumpf, S. A., and R. A. Longman. 2000. The ultimate consultant: Building long-term, exceptional value client relationships. *Career Development International* 5 (3): 124–134.

Sturdy, A. 2009. Popular critiques of consultancy and a politics of management learning. *Management Learning* 40 (4): 457–463

Troper, J., and Lopez, P. D. 2009. Empowering novice consultants: New ideas and structured approaches for consulting projects. *Consulting Psychology Journal: Practice and Research* 61 (4): 335–352. .

van Es, R. 1993. On being a consultant in business ethics. *Business Ethics: A European Review* 2 (4): 228–232.

Viola, J. J., and S. D. McMahon. 2010. Preparing for success. In *Consulting and Evaluation with Nonprofit and Community-Based Organizations*, eds. J. J. Viola and S. D. McMahon, 19–31. Sudbury, MA: Jones and Bartlett.

Webb, A. D. 1993. Consultants and their work: The learning dynamic. *New Directions for Adult and Continuing Education* 58 (2): 21–29.

Weiss, A. 2009. *Getting started in consulting*. 3rd ed. Hoboken, NJ: John Wiley & Sons.

14

"Work" and "Life": Crossing Boundaries of Time, Space, and Place

Janice Monk

As the chapters in this book demonstrate, opportunities to practice as a geographer today are many, varied, and widening. So too are the places where geographers work. They may be employed internationally, on development projects, or in the diplomatic service; they may make a career in a large corporation in a metropolitan area, work for local government in their state, or serve as a home-based consultant. Geographers are also a very diverse group of people. Think of those you know. They likely include women and men, single people or one of a dual-career couple (married or not, with a same or opposite sex partner), people with or without children. They may be foreign-born with older parents living in another part of the world. They may be in a commuting relationship.

When I was young, my father used to describe his colleagues as "living to work" or "working to live." That dichotomy no longer resonates with me or, I imagine, with many of you. We need, expect, and want to do both. We make friendships with colleagues where we work and across the profession. We engage in community life and desire time for recreation. So how do we bring these worlds together? Daily practices and policies in the workplace have changed over recent decades, especially with the rise of electronic communication. But how well do they parallel or support the extent of changes in other parts of life?

This chapter will examine some of the strategies professionals use to deal with the challenges of managing time and space in their personal and work lives and the choices available to them. The type of organization and its location have implications for combining "work" and "life" and advancing in a career. The first section will consider some general aspects of valuing and managing time. The second section will introduce examples of policies and modes of work that are relevant for managing the integration of work and life. The third will offer examples that go beyond the here and now as professionals aim to advance in their careers, whether remaining in the same workplace, moving to a new location in their own country, or engaging in international work. Finally, the chapter will suggest future challenges for individuals, employers, researchers, and policy makers if choices and opportunities are to be enhanced.

VALUING TIME, CONSIDERING SPACE

In a recent study of geographers working in business, government, and nonprofit organizations, respondents placed "time management" at the top of the list of skills they most often needed to practice in their work (Solem, Cheung, and Schlemper 2008). This finding raises important questions not only about how work is carried out, but also about how work and other parts of life come together. That theme is the subject of an extensive literature addressing how households at different points in their life courses and in various fields of work cope with work-life relations (Moen 2003; Kossek and Lambert 2005; Bianchi, Robinson, and Milkie 2006; Christensen and Schneider 2010). It is recognized that the hours people spend in paid work have been increasing as internationalization of economies, economic restructuring, and outsourcing shape business practices and as technological innovations have made it feasible for people to carry out their work responsibilities more or less at any time and in any place. While electronic communication has created new individual options, it places new demands and sources of stress on personal life. The world of work has moved from 9-5 to 24-7 (Burke and Michie 2002). Research indicates that American (and also British and Canadian) workers put in longer hours than those in other industrialized countries (Hardill 2002)[1]. The U.S. Bureau of Labor Statistics reported that in 2003, for example, 40 percent of male managers and 20 percent of female managers worked 49 or more hours per week (Kossek and Lambert 2005).

Many studies address how gender is implicated in the unequal division of family work. This research focuses mostly on married couples with children and how their choices vary by life stage, education, and the values they hold about egalitarian relationships. Though gender differences have become less pronounced as women increasingly have taken up careers, in general it appears that women still assume more of the domestic responsibilities than men in ways that may inhibit women's career advancement (see, for example, studies in Moen 2003). Much less is known about how other types of households, such as those of same-sex couples, manage, though some differences have been reported in the extent to which male or female couples rely on partners for social support (Mock and Cornelius 2003).

People may value time very differently from one context to another: for example, how we manage time at work may be different from how we think about time spent at home. Because of this distinction, a good starting point is to reflect on how you value and manage your time in different situations. What might your current approaches imply for the work context and for your life beyond the workplace? Here the literature on time management can offer some suggestions for analyzing your daily patterns, for example, by keeping a time log. Make a list, such as that in Table 14.1, of the broad set of tasks that you normally expect to do in a day. First, record your time use on three sample days in the week. Of course, we often multitask—parents may prepare a meal while supervising a child's homework, or you may run the laundry while doing e-mail. Next, record what you would consider your ideal allocation. Then assess how your ideal and actual compare. What changes might you want or be able to make? Stephen Covey (1989) has suggested that a good way to begin prioritizing your time is to identify activities by whether you would consider them as urgent/important; not urgent/important; urgent/not important; or not urgent/not important. Such a classification can help you to evaluate whether you are using time in ways you value.

As geographers, we know that activities take place in space as well as time. So how might the spaces in which you live and work affect the ways in which you spend time? Research carried out with urban households in the Netherlands might stimulate you to think about alternatives.

[1]Because work hours are measured differently in different settings, making comparisons is complex. Some indication can be gained from data compiled by the Organization of Economic Co-operation and Development. See http://stats.oecd.org/Index.aspx?DataSetCode=ANHRS (last accessed 16 July 2011).

TABLE 14.1 How Do You Spend Your Time?

Estimate the amount of time (hours or minutes) that you spend on the tasks listed below in a day during your work week. Repeat for three sample days. How does this use of time compare with what you would see as your ideal?

Paid work (you could break this down by key regular tasks)

Dealing with e-mail
Other discretionary online and phone activities (e.g., social networking, blogging, connecting with family and friends)

Commuting
Maintenance tasks (e.g., shopping, meal preparation, laundry)
Self-care (sleeping, eating, bathing)
Caring for others (e.g., partner, children, parents, or friends)

Exercise
Social/leisure (e.g., activities with friends, reading, watching TV)
Community and/or professional service
Multitasking

In one study, Hubers, Schwanen, and Dijst (2011) identified seventeen aspects of paid work and twenty-nine of domestic tasks that might be used as coping strategies to strive for work-life balance. They asked members of single- and dual-earner households which combinations of strategies they used and examined how these combinations were influenced by individual and household backgrounds. Work-related strategies involved individual approaches such as reducing the number of hours of paid work or getting a job closer to home; others depended on using material goods (for example, using a laptop to work "anywhere, anytime"). Among strategies for reducing time spent on domestic tasks, some involved individual behavior, such as sacrificing leisure time to caring tasks, others substituted external costs (such as buying take-away food or hiring house cleaning), and some involved use of social networks (such as sharing days of child-care with friends or relatives). This is not the place to review their detailed findings, though they show that while the combination of strategies adopted depended strongly on the presence of young children in the household it also reflected the characteristics of the home neighborhood and main workplace.

Related research by Schwanen, Ettema, and Timmermans (2007) found that sociodemographics remained critical. Women in their study still performed the bulk of the out-of-home household tasks such as shopping, but access to public transportation and characteristics of neighborhood land use patterns (such as higher densities) were associated with increased participation by men in out-of home household duties. What their research offers is ways of thinking about options and also what constraints might need to be addressed to accomplish a more equitable and satisfying balance of work and life. Other research on managing work and personal time highlights the importance of going beyond static approaches to take into account the increasing diversity of individual life paths, styles, and contexts (Moen and Sweet 2001). Your place of employment will also affect your choices and constraints. Geographers Schlemper and WinklerPrins (2009) offer suggestions for those in academic employment, but their ideas can be adapted to other settings.

Although the literature emphasizes managing domestic and work time, it is also important to think about other aspects of social time as they connect to "work" and the quality of life. A widely publicized book by Robert Putnam, *Bowling Alone* (2000), argued that Americans were increasingly disconnected from family, friends, and neighbors, thereby impoverishing community life. His views received criticism as well as support, but they do prompt reflection on the various ways in which work-life connections are of value in the community and in the world of work. In a study of married, midlife women in managerial positions, Roos, Trigg, and Hartman (2002) found that while sustaining commitments to family, these women saw work as a place for forming friendships, enhancing their sense of personal identities, and offering opportunities for engagement in the community service activities that they valued. This community work, in turn, enhanced their careers, developed their skills, and contributed to the goals of their companies. Such a model is illustrated in the profile of geographer Carmen Masó, who is active in community organizations and whose choice of a place to work also reflected her concerns about changes in her mother's health.

WORK POLICIES, CHOICES, AND PRACTICES

Just what options we have to manage our work and personal time reflect national policies and those of employers. Compared with most Western countries, for example, the United States lacks policies for mandated vacation time or paid maternity leave (Moen and Roehling 2005). The most comprehensive American legislation is the Family and Medical Leave Act (FMLA) (U.S. Department of Labor 2011) of 1993, which allows "eligible employees of covered employers to take unpaid, job-protected leave for specified family and medical reasons with continuation of group health insurance coverage under the same terms and conditions as if the employee had not taken leave." It allows twelve weeks of leave in a twelve-month period if an employee has a serious health condition that prevents performance of essential functions of his or her job; for childbirth and care of infants up to one year of age; and other family-related issues such as adoption of a child and care of a spouse, child, or parent who has a serious health condition. Key phrases here are "eligible employees" and "covered employers." The FMLA applies to "all public agencies, including state, local and federal employers, local education agencies (schools), and to private-sector employers who employed fifty or more employees in twenty or more work weeks in the current or preceding calendar year, including joint employers and successors of covered employers" (U.S. Department of Labor 2010). Thus if you are making a career in a smaller, private-owned business, the employer is not required to provide you with this support.

Federal agencies also offer alternative work schedules to allow employees to adjust their daily hours at work for such purposes as avoiding peak-hour traffic or using nonweekend time to accomplish personal business. The program requires that workdays consist of core hours, when all employees must be at work, and flexible hours within time "bands" set by their agencies when employees may choose their time of arrival and departure (U.S. Office of Personnel Management 2011a). The federal government also offers options for part-time work and job sharing, which may accommodate other needs and stages of life such as time to care for children or the elderly, or to take a phased retirement (U.S. Office of Personnel Management 2011b). There remain uncertainties and barriers about using such programs, however. Employees may think that being present ("face time") is important for relations with colleagues and also wonder how they

PROFILE 14.1

Carmen Masó, GIS Analyst, U.S. Environmental Protection Agency (Chicago, IL)

Carmen Masó is a GIS Analyst who works in Chicago as Coordinator for the Environmental Protection Agency Region 5. A Latina who grew up in that city, she had an early interest in meteorology, but when she started college at the University of Illinois in Champaign-Urbana, Carmen majored in geography because there was no undergraduate program in meteorology. She sustained her interest in meteorology by doing a volunteer internship with the National Weather Service. Next she went on to the University of Oklahoma for graduate work in meteorology, but while there became interested in remote sensing and GIS, so she switched to earn a master's degree in geography. Her thesis addressed the use of satellite imagery in investigating thunderstorms in the Central Plains region.

Carmen's combination of skills and interests led to her current career. Her first position was with the Environmental Protection Agency as a GIS Analyst in San Francisco where she worked from 1992 to 1999. When her mother in Chicago became ill, Carmen was able to transfer to the EPA office there. Her move illustrates one of the advantages of employment in the federal government—the possibilities of moves within an agency. It also shows that family considerations may prompt a change of location within a career.

Carmen's work with the EPA involves quite a lot of outreach to bring the EPA's programs and resources to people in other agencies, businesses, and the community. She is Leader of the Region 5 GIS Response Support Team for Emergency Response. One of her major assignments involved the development of interactive mapping applications that are used in carrying out environmental impact assessments. She has also trained regional staff on principles of environmental justice. Carmen's accomplishments have been recognized with a number of awards, among them an EPA Gold Medal for her contributions to the agency's enforcement work, and Bronze Medals for her work on Midwest floods and on the Indiana Harbor Project. She has also received awards for her efforts with the Hispanic Employee Program and with Women in Science and Engineering.

Beyond her workplace, Carmen is active in community service and professional organizations. She enjoys sharing her interests in geography with people of all ages, from involvement with the State Geography Bee to volunteering at the Adler Planetarium, which selected her as Volunteer of the Year. She is a Board Member of the Chicago Geographic Society and participates in the Windy City chapter of Federally Employed Women. Carmen also belongs to the Illinois GIS Association, the Urban Regional Information System Association (URISA) and the AAG. She presents at their professional conferences. To develop the speaking skills that are important in both her paid and volunteer work, Carmen became a member and leader in the local EPA Toastmasters International club, an organization whose members practice and get feedback on their presentations. Carmen's "work" and "life" have certainly come together.

—JANICE MONK

will be evaluated by supervisors if part-time work tends to be seen as having limited opportunities for promotion (Blunsdon et al. 2006).

You will find that options and benefits in the private sector vary considerably, so when searching for a position it is important to learn what you can about details and their implications for your lifestyle. Annual paid vacation time in the early years of a position may be only one week. It may, however, increase as you remain with a company. Other issues to check are coverage for sick days and health insurance (and if this includes options for partner or family coverage), and whether the employer contributes to a retirement plan. Is there support for professional development, for example, if you wanted to participate in the annual Applied Geography Conferences or the annual meeting of the Association of American Geographers? Would your conference and travel costs be paid for wholly or partially? These will likely not be issues you would want to address on your first inquiry or interview, but larger organizations may well have information online. The geospatial employer, Esri, for example, provides online information on the range of benefits it offers for personal and family well-being such as life insurance and educational assistance (Esri 2010). Using networking to find out about work conditions and cultures can be helpful in how to assess your options (see Tina Cary's Chapter 5 on professional networking for more on this topic). Background on laws affecting flexibility, including some state leave laws, can be found on websites such as Georgetown Law's (2011) Workplace Flexibility 2010 project. The organization WorkOptions (2011) offers guidance for preparing individual proposals to request flexible work options such as telecommuting, compressed workweeks, or job sharing. The work-life literature suggests that while employees benefit most from flexible work programs, there are also benefits to employers in terms of attracting and retaining well-qualified staff, and reducing absenteeism (Valcour and Batt 2003).

Numerous studies have examined changing trends over time and the involvement of different groups of workers in available work-life options. Table 14.2 illustrates some of the recent

TABLE 14.2 Selected findings from *Work-life: Prevalence, utilization, and benefits* (Catalyst 2011)

- 44% of senior-level women compared to 36% of men in corporate leadership used flexible arrival and departure time
- 13% of women compared to 12% of men in corporate leadership telecommuted/worked from home
- 91% of women and 94% of men agreed they could be flexible in schedules when they had a family emergency or personal matter, but only 15% of women and 20% of men agreed they could use a flexible work arrangement without jeopardizing their career advancement
- 79% of employers allowed "some employees to periodically change start and quit times within some range of hours, but the number dropped to 37% when asked about "most or all employees"
- 23% allowed "some employees" to work part of the workweek at home occasionally, but the number dropped to 1% when asked about "all or most" employees
- Flexible schedules were more common among white workers (28.7%) and Asian workers (27.4%) than among black (19.7%) or Latino (18.4%)
- Between 1977 and 2008, dual-career men reported increasing work-life conflict from 34 to 45%, and women from 34 to 39%
- 50% of companies in the *2008 National Study of Employers* reported that they trained supervisors in responding to family needs of employees
- 67% of employees at organizations with high levels of flexibility report high levels of job satisfaction compared to 23% of employees at organizations with low levels

findings summarized in a report prepared by Catalyst, an organization that focuses on people in professional and managerial positions. Research by the Families and Work Institute (2011) also offers a wealth of information on changing practices and attitudes. One of the interesting findings is of generational change in satisfactions. Compared to other age groups, younger professionals (the "Millennials") less often than other age groups feel that choosing flexible starting and quitting times will meet personal needs (47 percent compared to 83 percent of all employees), even though they officially have the same access to flexibility. Younger workers more often than their predecessors also report that they do not want jobs with more responsibility because of long hours and limited flexibility. The Catalyst report suggests that employers may need to take into account that younger workers, having grown up in the high-tech world, may prefer approaches to flexibility other than those of flexible starting and quitting times (Matos and Galinsky 2011). They may prefer more support for telecommuting so that they can exercise greater personal control over where and when to work. This research also implies that as you enter a professional position, you should explore your personal expectations in relation to time and place of work and what those might mean in a given position.

As an option, telecommuting may be part-time (e.g., one or more days per week) or may be the principal means by which a person carries out his or her work. Many people do a form of "telecommuting" (or after hours) work at home—doing business e-mail, preparing reports and presentations, and so on. Most of you will be familiar with that phenomenon from your student experiences. What are desirable ways you can use technology to work other than in the space provided by an employer? The extent and nature of such work cannot readily be documented, though the number of U.S. employees who worked remotely at least one day per month was reported to be 17.2 million in 2008, an increase of 39 percent from two years earlier; a shift away from full-time telework to occasional telework has also been noted (World at Work 2009a). Those most willing to telecommute are male college graduates in higher income households who are under age 55, and mostly around 40 years of age. Most often they worked from their home, though some telework may also be carried out from a car (though it is hoped not while driving!) or from a customer's place of business (World at Work 2009b). Of studies conducted by the Families and Work Institute and reported by Catalyst (2011), one indicated that among senior-level employees, 13 percent of men and 12 percent of women telecommuted or worked from home. Another reported that while 23 percent of employers allowed "some employees" to work part of the workweek at home occasionally, the number dropped to 1 percent when they were asked about "all or most" employees.

As a mode of work, telecommuting raises a number of questions about time and space. What home spaces are used for work? Is there a designated personal office or reliance on shared living space? Is work time restricted to a set range of hours? How are boundaries maintained, and how are leakages of time and space managed? Who owns or pays for equipment and services and their maintenance? Do those who work primarily as telecommuters include face-to-face relationships with fellow professionals, and if so, how? At what point in a career does a professional enter into this kind of work arrangement and why? Limited research addresses these questions. A review by Chesley, Moen, and Shore (2003) noted that among married couples with children, both fathers and mothers have reported that managing home work and family life presented difficulties, and that some children whose fathers worked from home thought that their father had difficulty focusing on them. Research by Sullivan and Lewis (2001) and Johnson, Andrey, and Shaw (2007) offers case studies illustrating ways in which women and men manage teleworking. For geographer Jeff Young (see the accompanying profile), who began telecommuting midcareer, it has offered a positive approach for combining his work, his wife's employment outside the home, and the needs of school-aged children.

PROFILE 14.2

Jeff Young, Business Development Manager, LizardTech (Seattle, Washington)

In the early years of his career, Jeff Young was very mobile both within the United States and internationally. But for some years now he has worked from his home in Colorado, though with companies located elsewhere—currently in Seattle and before that for companies in Atlanta and Boca Raton. Jeff's early positions involved technical tasks. After he earned his M.A. at Arizona State University, his first positions drew on his technical skills in remote sensing and environmental geography to undertake such projects as floodplain mapping, site selection, and demographic analysis. Now he is business development manager for geospatial solutions with LizardTech, a software company that specializes in geospatial image compression. Celartem, LizardTech's parent company, is based in Japan and has experience in handling off-site management relations. It is quite supportive of arrangements such as Jeff's.

Family priorities have been a key element in the transitions Jeff has made. Once he married and had a son and daughter, he and his wife became disenchanted with a lifestyle in which he might be out of the country for extended periods. His wife is a professional who can obtain stable employment locally. They decided that Jeff's skills and connections presented them the options of working from their home base. Jeff has a breadth of experience and has been active in professional organizations that have given him good networks. So he is well suited for a position that involves business development by identifying and supporting potential governmental and private users of his company's geospatial software. He describes his position as being "something like a diplomat," one who establishes personal connections and builds relationships of trust.

So how does Jeff manage in his day-to-day life? Working from home has many advantages for both the employer and employee, particularly in allowing efficiencies in time management where multiple time zones are a factor in daily communication. He saves substantial daily time that would be eaten up by commuting. He can engage in family life—he is involved in the day-to-day worlds of his son and daughter—and he shares household tasks with his wife, who now works regular hours as a school nurse. Jeff considers that managing his career requires that one be well organized and have a daily plan. He values the lack of distractions at home that could occur in other work spaces, yet he can also take breaks to think and plan while doing some household tasks. Of course, Jeff doesn't really work in isolation. He goes to corporate headquarters for face-to-face contact once a quarter and may increase that contact into a four- to six-week cycle. He maintains regular contact with management and relevant staff on conference phone calls as well as electronically. The company supports his communications. He attends relevant conferences where he establishes and maintains relationships with clients. Jeff stresses that to work as he does one needs to be self-motivated and to earn the trust of his direct manager and his employer in general. As role models, he credits his father, who still practices as a self-employed architect, and his mother, who served as a lifelong NGO volunteer. Both showed the way by instilling in Jeff a strong work ethic and a sense of purpose.

—JANICE MONK

MOVING UP, MOVING ON, GOING FURTHER AFIELD

At different points in your career or stages in life, you will likely consider making changes. You may want a promotion or a change of scene, or you may be making a change in your personal life. You may have started your career in one sector, for example in higher education, but think that you would like to try other settings (see Chapter 3 by Joy Adams on switching sectors and making career transitions). Because the world of work is a dynamic one, especially in an era of globalization, multi-sited corporations, transformations and uncertainties in national economies, and changing technologies, retaining a position or advancing in your career may require short- or long-term relocation. Dual-career households are commonplace, so you may well have to consider the implications of moving on for two people's careers, or for children's care and education, or the sometimes even more complicated issues of "blended" and extended families. If you came to the United States as an international student, there will also be the question of whether to return home, and if not, how immigration regulations impinge on your career opportunities. So what are your personal options and constraints? How are your identity and your work intertwined? What would it mean to be in a commuter relationship? Do you have a long-term career strategy? Do you see yourself on the fast track? Are you willing to try self-employment (see Chapter 13 by Susanne Moser and Angela Donelson for more information on this topic). Should you move on or stay where you are?

As noted in Chapters 6 and 7, positions in federal, state, and local government agencies generally offer job security, health and retirement benefits, and more predictable career ladders than positions in other sectors. Still, in some federal government agencies, advancement may involve moving to another location; frequent moving is to be expected in some fields such as the diplomatic service or in development agencies. In private business today it is not unusual to shift to another company or transfer (or have the option to transfer) to another location in the home country or internationally in order to advance. Short-term contractual work is also becoming more common.

This chapter cannot address all the options and contingencies. Although some research reports that there has been a shift away from prioritizing the husband's career, gender differences still exist. In one study, twice as many men as women reported facing a career opportunity that would also have required their spouse to move or change jobs, with almost half of the men reporting having taken that choice compared to one-third of the women (Pixley and Moen 2003). Geographer Irene Hardill's (2002) research with women and men professionals in Britain, Canada, and the United States explores some implications and experiences of commuter relationships among dual-career households. She found that when couples moved, the woman was often the "trailing partner." She was likely to face downgrading in her career, with likely consequences for household income and for her longer-term advancement.

What are some of the options and conditions to consider? Are there options for one partner to telework? Does the corporation offer relocation assistance? What if you decided to have a commuting relationship and keep up two households? What are the financial and emotional costs? Consider the example of a Canadian academic woman whose husband works internationally as a development consultant:

> Given the experience, four weeks is something I can accept, which is fine. More than that becomes much more difficult because you have to get installed in a new situation. It's more difficult when he comes back in a sense too. Right now it is three weeks . . . it's just like a weekend for us. But last time was eight weeks and then you

> have to think in terms of being a single parent for two months, which is very different. (Hardill 2002, 29)

She continued that their five-year-old daughter also found it difficult in the early stages of her father's absence, though she got used to the situation in about two weeks and could do well.

If assignments are in remote places, communication may not be possible for extended periods. If location and time differences permit, however, regular phone calls or use of Voice over Internet Protocol (VoIP) services and applications such as Skype may help with communication. Costs of long-distance weekend commuting travel (including by air) may be supported by the employer (Hardill 2002). Even so, the lifestyle of the absent partner can be unpleasant, at times requiring one to live in subpar hotels or apartments, while social contacts may be limited to co-workers. For those who come to the United States as international students and then seek to remain for their professional careers, the challenges will be compounded (see profile of the Chinese couple, Mei and Jun).

The increasing phenomenon in international work as outsourcing to contractors replaces longer term government assignments (Hindman 2011) widens the group of professionals who have to manage extended separations from friends and families. Those employed by national governments or large international organizations, for example in the diplomatic service, may expect frequent moves, but compared with contract employees they have the advantages of longer assignments, urban living, and support services such as those offered to employees of the World Bank through its Global Mobility and Work Life Programs (2011). Still, taking up an international career may mean experiencing life in an "ex-pat bubble" and facing issues of whether children can be placed in international schools or whether to send them back to the home country for education. There are also difficulties in finding professional employment locally for spouses (Coles and Fechter 2008).

CONCLUDING THOUGHTS

One implication of this chapter is that it is important to consider your personal priorities, strengths, and concerns as you assess your career opportunities and goals, and to revisit how you think about them periodically throughout your life, especially as your personal relationships change and your decisions involve other people close to you. Issues that may seem paramount now may be less so later. Factors that may at present seem immaterial to your career may become very important later. It is important also to research work-life policies and options in places/sectors of employment you are thinking about and ask questions as needed along the way.

Overall, the demographics of the workforce have probably changed faster over the past generation than have workplace policies and cultures. Dual-income families, single-parent families, single-income individuals living alone or with partners, unmarried couples with children, blended families, and other personal and family arrangements have become common, perhaps even the norm. The workplace still tends to be geared toward the traditional nuclear family in which personal and family responsibilities can be shared by partners, whether to take care of each other, their children, or aging or ill family members. But this doesn't mean that change is impossible. Policies and practices are dynamic, and an array of researchers and organizations are devoting attention to studying the issues and advocating for changes that would better serve current realities. So too do you have the possibilities for innovating in your own life and for advocating for alternative contexts in which you work that would benefit both employees and employers.

PROFILE 14.3

Mei and Jun

Mei and Jun[1] are an early-career couple, but each lives in a different city. They first met as graduate students at the same U.S. university where Mei was doing doctoral studies in geography and Jun in environmental sciences. At the time, in their home country of China, the cultural trend was to go abroad for graduate study. Despite the challenges of studying and working in English, their choice now is to make their careers in the United States, partly because their interests and disciplinary orientations have changed over the years and partly because their networks have shifted so that they are better suited to developing careers here. But they also face obstacles that are unique to those who enter the country on student visas and wish to remain after graduation.

Mei waited to seek a position until she completed her doctorate and was able to secure a tenure track academic appointment. She then started the process of seeking a work visa[2] that would give her eligibility for three years of employment and a possible three-year extension. Meanwhile, Mei started working on the permanent resident (Green Card) petition, a lengthy, chancy, and expensive process that creates stress during the years one is trying to get professionally established. Her university has an office to assist the employing department and the scholar with securing the H1-B (work visa) or EB1/EB2 (for the Green Card) for permanent residence. But the processes are complex and time consuming, with a successful outcome not guaranteed.

Before completing his dissertation, Jun took advantage of a legal option for international students to gain work experience. This began the couple's period of living in separate communities. After a year of that experience and completion of his doctorate, Jun sought a position in the private sector, drawing mainly on the methodological skills he had acquired in his graduate education. Such appointments require that the employer be willing to sponsor the job applicant and to manage the legal work of applying for the visa status. Finding such an employer limits where one can work beyond the opportunities open to graduates who are citizens. The position Jun found meant living very far from Mei. They could only commute by plane and manage to be together in the summer when Mei was not teaching. Otherwise communication was by phone, e-mail, or online messenger.

After a year in this situation, Jun began to look for other positions that would be closer to Mei. The choices were constrained by an array of barriers—when the pay was better, for example, the distance still was too great in terms of time, travel, and living costs, and it still meant maintaining two residences. Eventually, Jun found a private sector position that was only about 2 hours from where Mei works, though at a substantially lower salary than in the first job. The location allows them to get together on weekends at Mei's home, while Jun lives in a small apartment shared with a coworker. Jun has since found another position in the same city that pays better. Work options remain limited by the need to find an employer who will take on the legal obligations and costs of supporting an application for the required work visa.

In today's tight labor market, barriers to obtaining a work visa have risen substantially. Employers have to demonstrate that the applicant is better qualified to hold the job than a U.S. citizen. Furthermore, because work visas can only be held for a limited time, both Mei and Jun need to seek permanent residency status. Success in this application is a time-consuming, expensive, and uncertain process.

Mei and Jun's situation is not unusual. They have to live with uncertainties and be resilient. While managing lives away from their original families, restricted in their opportunities to leave the United States even for short periods, they gain support from each other and from local friends and colleagues who may have similar backgrounds.

—JANICE MONK

[1]These are pseudonyms.

[2]Information on the types of visas and the complicated application processes for those seeking permission for permanent employment can be found at http://www.uscis.gov (last accessed July 10, 2011)

REFERENCES

Bianchi, S. M., J. P. Robinson, and M. A. Milkie. 2006. *Changing rhythms of American family life.* New York, NY: Russell Sage Foundation.

Blunsdon, B., P. Blyton, K. Reed, and A. Dastmalchian. 2006. Introduction: Work, life, and the work-life issue. In *Work-life integration: International perspectives on the balancing of multiple roles,* eds. P. Blyton, B. Blunsdon, K. Reed, and A. Dastmalchian, 1–16. New York, NY: Palgrave Macmillan.

Burke, N., and S. Michie. 2002. New directions for studying gender work stress, and health. In *Gender, work stress and health,* eds. D. L. Nelson and R. J. Burke, 229–242. Washington, DC: American Psychological Association.

Catalyst. 2011. *Work-life: Prevalence, utilization and benefits.* http://www.catalyst.org/publication/238/work-life-prevalence-utilization-and-benefits (last accessed 5 July 2011).

Chesley, N., P. Moen, and R. Shore. 2003. The new technology climate. In *It's about time: Couples and careers,* ed. P. Moen, 220–241. Ithaca, NY: Cornell University Press.

Christensen, K., and B. Schneider. 2010. *Workplace flexibility: Realigning twentieth century jobs for a twenty-first century workforce.* Ithaca, NY: Cornell University Press.

Coles, A., and A-M. Fechter, eds. 2008. *Gender and family among transnational professionals.* New York, NY: Routledge.

Covey, S. R. 1989. *The seven habits of highly effective people.* New York, NY: Simon and Schuster.

Esri, Inc. 2010. *Esri employee benefits.* http://www.esri.com/library/brochures/pdfs/esri-employee-benefits.pdf (last accessed 5 July 2011).

Families and Work Institute. 2011. Research and publications. http://familiesandwork.org/site/research/main.html (accessed 5 July 2011).

Fechter, A-M. 2011. Anybody at home? The inhabitants of Aidland. In *Inside the everyday lives of development workers: The challenges and futures of Aidland,* eds., A-M. Fechter and H. Hindman, 131–149. Sterling, VA: Kumarian Press.

Georgetown Law. 2011. Workplace flexibility 2010. http://workplaceflexibility2010.org/ (last accessed 5 July 2011).

Hardill, I. 2002. *Gender, migration and the dual career household.* London, UK: Routledge.

Hindman, H. 2011. The hollowing out of Aidland: Subcontracting and the new development family in Nepal. In *Inside the everyday lives of development workers: The challenges and futures of Aidland,* eds., A-M. Fechter and H. Hindman, 169–191. Sterling, VA: Kumarian Press.

Hubers, C., T. Schwanen, and M. Dijst. 2011. Coordinating everyday life in The Netherlands: A holistic quantitative approach to the analysis of ICT-related and other work-life balance strategies. *Geografiska Annaler Series B* 93 (1): 57–80.

Johnson, L.C., J. Andrey, and S. M. Shaw. 2007. Mr. Dithers comes to dinner: Telework and the merging of women's work and home domains in Canada. *Gender, Place and Culture* 14 (2): 141–161.

Kossek, E. E., and S. J. Lambert, eds. 2005. *Work and Life Integration: Organizational, Cultural, and Individual Perspectives.* Mahwah, NJ: Lawrence Erlbaum Associates Publishers.

Matos, K., and E. Galinsky. 2011. *Workplace flexibility among professional employees.* New York, NY: Families and Work Institute. http://familiesandwork.org/site/research/reports/WorkFlexAndProfessionals.pdf (last accessed 5 July 2011).

Mock, S. E., and S. Cornelius. 2003. The case of same-sex couples. In *It's about time: Couples and careers,* ed. P. Moen, 275–287. Ithaca, NY: Cornell University Press.

Moen, P., ed. 2003. *It's about time: Couples and careers.* Ithaca, NY: Cornell University Press.

Moen, P., and P. Roehling. 2005. *The career mystique: Cracks in the American dream.* Lanham, MD: Rowman and Littlefield.

Moen, P., and S. Sweet. 2001. From work-family to flexible careers. *Community, Work and Family* 7 (2): 209–226.

Pixley, J. E., and P. Moen. 2003. Prioritizing careers. In *It's about time: Couples and careers,* ed. P. Moen, 183–200. Ithaca, NY: Cornell University Press.

Putnam, R. D. 2000. *Bowling alone: The collapse and revival of American community.* New York, NY: Simon and Schuster.

Roos, P. A., M. K. Trigg, and M. S. Hartman. 2002. Changing families/changing communities. *Community, Work and Family* 9 (2): 197–224.

Schlemper, B., and A.M.G.A. WinklerPrins. 2009. Balancing personal and professional lives. In *Aspiring academics: A resource book for graduate students and early Career faculty,* 42–51. Upper Saddle River, NJ: Pearson Prentice Hall.

Schwanen, T., D. Ettema, and H. Timmermans. 2007. If you pick up the children, I'll do the groceries: Spatial differences in between-partner interactions in out-of-home household activities. *Environment and Planning A* 39 (11): 2754–2773.

Solem, M., I. Cheung, and M. B. Schlemper. 2008. Skills in professional geography. *The Professional Geographer* 60 (3): 356–373.

Sullivan, C., and S. Lewis, 2001. Home-based work, gender, and the synchronization of work and family perspectives of teleworkers and their families. *Gender, Work and Organization* 8 (2): 123–145.

U.S. Department of Labor. 2010. Fact sheet #28: The Family and Medical Leave Act of 1993. http://www.dol.gov/whd/regs/compliance/whdfs28.htm (last accessed 1 July 2011).

U.S. Department of Labor. 2011. Family and Medical Leave Act. http://www.dol.gov/whd/fmla/ (last accessed July 1, 2011).

U.S. Office of Personnel Management. 2011a. Alternative work schedules: Flexible work schedules. http://www.opm.gov/oca/worksch/html/awsfws.asp (last accessed 5 July 2011).

U.S. Office of Personnel Management. 2011b. Part-time work and job sharing. http://www.opm.gov/employment_and_benefits/worklife/workplaceflexibilities/jobshare/ (last accessed 5 July 2011).

Valcour, M., and R. Batt. 2003. Work-life integration: Challenges and organizational responses. In *It's about time: Couples and careers*, ed. P. Moen, 310–331. Ithaca, NY: Cornell University Press.

WorkOptions. 2011. Are you squeezed for time? Here's the easiest way for you to ask for flexible work and get management approval. http://workoptions.com/ (last accessed 2 July 2011).

WorldatWork. 2009a. Telework revs up as more employers offer work flexibility. http://www.workingfromanywhere.org.news.pr021609.html (last accessed 5 July 2011).

WorldatWork. 2009b. Telework Trendlines. 2009. http://www.worldatwork.org/waw/adimLink?id=31115 (last accessed 5 July 2011).

World Bank. 2011. Global mobility and work life programs. http://go.worldbank.org/LJMEIZ4G10 (last accessed 5 July 2011).

15

Practical Ethics for Professional Geographers

Francis Harvey

Without ethics, everything happens as if we were all passengers on a big truck without a driver; and the truck is driving faster and faster, without us knowing where.

—Jacques-Yves Cousteau

Ethics matters for geographers. This means considering issues of professional responsibility and the ethical dimensions of work-related activities. More than deciding "rights from wrongs," professional ethics involves thinking about the impacts, consequences, and moral implications of our work. Cousteau's analogy is nicely chosen—without ethics, our work and careers can careen out of control. Ethics provides direction and guidance and helps us stay on the road. For professional geographers, ethical issues are paramount. We have a responsibility to ourselves, to our families and colleagues, and to society for our actions. Ethical issues can arise from even routine tasks: handling the confidentiality of clients and spatial data; resolving conflicts of interest; managing relationships and accountability with stakeholders on controversial projects; determining harm versus benefit in consulting; or assuring data integrity. As Cousteau implies, geographers must be willing to take the wheel of the truck and get passengers to their destination safely.

In this chapter, you will find out more about the practical aspects of ethics for geographers. Because the role of ethics for professional geographers is pervasive, this chapter will introduce a seven-step approach to ethical decision making that is broadly applicable and can be used to help evaluate ethical issues involving mapping, statistics, project management, and many other aspects of professional geography.

ENGAGEMENT WITH ETHICS

Professional geographers encounter many examples of ethical issues. Sometimes these issues arise from inadequate knowledge, insufficient expertise, or lack of thoroughness. These situations can result in compromised results, wrong conclusions, and bad decisions, all of which are emphasized in Tom Dwyer's profile where he reflects on his work as a consultant and appraiser. These issues are so important that a number of professional associations have developed statements or codes of professional conduct. The Association of American Geographers (AAG 2009) developed a statement on professional ethics fifteen years ago and has revised it periodically since. It should be among the first stops for students and practicing geographers, as well as the statements on professional ethics developed by other professional associations.

An example is the GIS code of ethics that is included among the requirements for certification as a GIS Professional (GISP) by the GIS Certification Institute (2011), a program that began accepting applications in 2004. Another relevant example is the engagement with ethics in the *GIS&T Body of Knowledge* (DiBiase et al. 2006). It highlights such ethical and legal issues as privacy, access, and intellectual property, and it is included among the "core" units that UCGIS recommends as part of every geospatial certificate and degree program curriculum. Similar statements of professional ethics can be found in neighboring fields such as the American Planning Association (APA 2009).

APPLIED ETHICS AND PROFESSIONAL PEDAGOGY

To dispel misunderstandings, it's worthwhile to begin with a few words about applied ethics and their relationship to professional geography. For purposes of this chapter, the term *applied ethics* refers to an area of philosophy that has received considerable attention in recent years. It approaches ethics with the idea that philosophy is relevant to the general public and engages issues relevant to day-to-day life. Much like professional ethics, applied ethics considers standards of ethical and moral behavior in commonplace situations (Quinn 2011).

Two things provide applied ethics with a good fit for professional geographers. First, as already mentioned, is an emphasis on engaging actual issues and being relevant to a larger public. Second is the emphasis in applied ethics on developing an empirical foundation for ethics. For example, philosopher Kwame Appiah (2008, 22) writes:

> Morality is practical. In the end it is about what to do and what to feel; how to respond to our own and world's demands. And to apply norms, we must understand the empirical contexts in which we are applying them. No one denies that, in applying norms, you will need to know what, as an empirical matter, the effects of what you do will be on others.

What Appiah is saying is of enormous importance for thinking about professional geography's ethics. Instead of looking for ethical theories and principles to explain choices, applied ethics involves considering both concepts and the actual situations and consequences. In other words, ethical engagement involves clear connection to examples.

A professional geographer working on a mapping project using Census data needs to consider not only the immediate project but also the larger situation. Mapping crime data and ethnic groups may be insightful at some level, but in a broader context what hidden assumptions should be brought to the surface? Often, professional geographers face conflicts of interest. What responsibility does a professional geographer have toward the funding body, a supervisor, colleagues, or other groups that may potentially be impacted by

PROFILE 15.1

Tom Dwyer, Principal, Dutch Hill Consulting (Poughkeepsie, New York)

Running any small business takes a lot of work. But managing your own single-person consulting firm can be especially demanding. "Every day has to be efficient," says Tom Dwyer of Dutch Hill Consulting, Inc. "Working for a small company means you must be constantly productive."

Tom's job involves estimating the value of commercial properties and large parcels of land. Because his work is often used to support litigation, many of his clients are attorneys, for whom he generates appraisal reports running anywhere from 40 to over 100 pages in length. Since a multitude of factors can potentially make an enormous difference in the value of a given property, Tom's **research is essential in informing his clients' decisions**. "Even a simple mistake can ruin the credibility of an appraisal and cause the litigation to turn in favor of the opposing party," he says.

When the results of one's work can mean losses or gains of thousands or even millions of dollars to a client, **self-confidence and strength of character** are essential personal attributes. Appraisers must obtain proprietary and sensitive data from individuals who can be uncooperative and abusive, and they frequently come under intense pressure to slant the results of their research. Tom has occasionally experienced verbal coercion, withholding of payment, and threats of legal action to attempt to influence his analyses. Despite these pressures, he has learned to "stick to his guns." In addition to considering the long-term gains for his clients (for example, is the cost of possible litigation worth any fees that might be recovered?), Tom also must be mindful of the **personal and professional costs of unethical practices**, which could result in the loss of his license or even jail time. In addition, referrals are key to maintaining and growing his business, so establishing trust with his clients and a solid reputation within the field are critical to his continued success. "By doing my job properly, I can have confidence in my work," Tom explains, adding that his rigorous approach means that he is able to defend his analysis and the final results when a client disagrees with an appraisal.

Commercial real estate appraisal is a licensed profession that mandates the completion of several core courses and at least two years of work experience under another certified appraiser, along with continuing education requirements. Tom notes that an advantage of working with a supervisor is the opportunity to **learn about potential ethical issues** early in one's career and to cultivate the strength needed to address them by observing how an experienced professional deals with such challenges. He also credits his geography education with honing the research, data analysis, and writing skills that are essential in his work; he strongly recommends coursework in economic geography and demography for aspiring appraisers. "I am frequently amazed when I see colleagues struggling over questions that could so easily be answered through spatial analysis," he observes.

Real estate appraisal entails understanding how numerous characteristics of a property such as zoning regulations, wetland and floodplain status, and demographic changes come together to influence its value; therefore, it is "very much a problem of geographic analysis." Tom's bachelor's and master's degrees in geography have proven invaluable assets, along with his previous professional experience. "The most important thing a consultant can have before going into business is a **depth of experience and knowledge** about the field," he notes. "This can only come through hands-on experience." Before launching his own business, he was employed by different companies performing retail site selection, shopping center research, real estate appraisal, and real estate data analysis. These positions provided him with the expertise, knowledge, and confidence needed to venture out on his own.

—MARK REVELL AND JOY ADAMS

the publication of this analysis and related maps? An ethnic group may consider the results of the work as an example of racial profiling and discredit the analysis and perhaps a larger project it is part of.

Appiah's statement that morality is practical is also relevant for considering morality in professional geography. Practical morality involves considering the situation in question, which is of course a key issue in applied ethics. Before getting to the specifics in the next section, remember that discovery of ethical issues and the deliberation of these ethical issues connect applied ethics to professional geography. First, we have the connection just mentioned between morality and professional ethics arising in an emphasis on considering the specifics of each situation. Second, applied ethical deliberation also reflects on the perspective and assumptions of different actors and ourselves in analyzing a situation—what feminists call standpoint (Walker 1998). Third, professional ethics focuses on considerations of the empirical situation.

While professional geographers often find ethical issues in hindsight after having made decisions in a short time frame, by engaging ethics in their early career training, geographers can develop a firmer foundation for such decisions. Professional geography pedagogy can benefit from connecting examples in the classroom to specific situations by using guest speakers, field trips, or role-playing exercises. These teaching methods can also help reduce differences between students with work experience and those with little or no professional work experience.

ETHICAL ANALYSIS

When teaching professional ethics, case studies can offer instructive scenarios that challenge students to analyze ethical problems and to identify reasoned solutions. Case studies also make it possible for professionals to connect their practical experiences to the discovery and deliberation of ethical issues. The "case method" is a common pedagogical technique for strengthening the moral reasoning skills of students in business, medicine, law, engineering, and computer and information science (Davis 1999; Keefer and Ashley 2001; Quinn 2006). Professional ethicists recommend that a framework guide the work with case analyses.

Discovery and deliberation in this section draw heavily on Davis's (1999) "seven-step guide to ethical decision making" (Table 15.1). Similar ethical analysis models have been suggested by Keefer and Ashley (2001) and others. The connection to scenarios in case study examples used with the seven-step method also allows the approach to be useful for engaging professionals.

EXAMPLES

The following two examples show how Davis's seven-step approach can be applied to develop responses to ethical problems. Both cases are available at the gisprofessionalethics.org website. The first case involves wetlands assessment and mapping. A more detailed description of this case and the analysis is available in DiBiase et al. (2009), from which the description of this case is taken. The second case, outlined in this chapter only, considers the commonplace issue of whether protected data should be shared with others. Other cases involving ethical issues from the introduction follow.

TABLE 15.1 Davis's (1999) seven-step guide to ethical decision making

Step 1. State problem. For example, "there's something about this decision that makes me uncomfortable," or "do I have a conflict of interest?"

Step 2. Check facts. Many problems disappear upon closer examination of the situation, while others change radically.

Step 3: Identify relevant factors. For example, persons involved, laws, professional code, other practical constraints.

Step 4: Develop list of options. Be imaginative, try to avoid "dilemma"; not "yes" or "no" but whom to go to, what to say.

Step 5: Test options. Use such tests as the following: Harm test: Does this option do less harm than alternatives? Publicity test: Would I want my choice of this option published in the newspaper? Defensibility test: Could I defend my choice of an option before a congressional committee or committee of peers? Reversibility test: Would I still think the choice of this option good if I were adversely affected by it? Colleague test: What do my colleagues say when I describe my problem and suggest this option as my solution? Professional test: What might my profession's governing body or ethics committee say about this option? Organization test: What does the company's ethics officer or legal counsel say about this?

Step 6: Make a choice based on steps 1–5.

Step 7: Review steps 1–6. What could you do to make it less likely that you would have to make such a decision again? Are there any precautions you can take as an individual (for example, announce your policy on question, change job, etc.)? Is there any way to have more support next time? Is there any way to change the organization (for example, suggest policy change at the next departmental meeting)?

Case Study: To Map Wetlands or Not To Map Wetlands?

Kelly is a geographic information systems (GIS) analyst and owner of a small environmental consulting firm that specializes in wetlands assessment and mapping. She has recently begun work on a project commissioned by the state of Oregon to identify estuarine areas on the Pacific Coast and to prioritize them for conservation and restoration. The contract represents a great opportunity and could lead to similar and even more lucrative contracts in Oregon and elsewhere.

The contract requires Kelly and her team to follow the client agency's established protocol for mapping tidal wetlands. The protocol involves several existing data sources. One is a digital map of probable tidal wetlands in the area (Scranton 2004). The protocol allows removal of polygons from this dataset if aerial photography interpretation, field visits, and other ancillary data suggest these do not represent actual tidal wetlands. In addition, areas may be added to the tidal wetlands dataset after field inspection if these areas are already identified and mapped in the National Wetlands Inventory (NWI) database. However, because the protocol is designed to be repeatable and usable by many people who may not have a background in wetland delineation, it does not include methods for adding new wetlands to the database that aren't already mapped in the NWI.

During Kelly's visits to sites of previously identified tidal wetlands, she finds evidence of additional wetlands that aren't mapped in either data source. The client agency's protocol doesn't accommodate the soil sampling needed to confirm Kelly's hypothesis. Neither does her project budget and schedule of deliverables provide the money or time needed to perform the extra work. She knows she cannot devote unbillable hours to the tasks either, since the project budget is barely adequate for the scope of work. There seems to be no way to verify with certainty whether these areas are wetlands. Leaving the sites out of her wetland mapping products could result in excluding important estuarine resources from conservation and restoration plans and, moreover, could reduce the overall efficacy of the agency's wetland conservation program.

These mapping-related issues are one side of the coin. Deciding to collect data that enhances conservation and restoration plans could violate the methodology of the protocol and could threaten the perceived integrity of her work. This could, in turn, cause her firm to lose money, which a small company cannot afford to do.

Kelly is in a difficult position for several reasons. The seven-step approach helps get a firmer grasp of the issues and possible approaches to resolving the situation.

Analysis Following the Seven-Step Approach

The following presents the discovery and deliberation involved in using the seven-step approach to analyze these ethical issues (Table 15.1).

Step 1: State problem. Analysis begins with a discovery of the ethical issues. This case involves a number of potential ethical issues related to the conflict of Kelly, an independent consultant with much experience, between her sense of environmental responsibility and her professional ambition, among them: Can field checking data adequately protect coastal wetlands? Would completing the contract following the scope of work reflect professional standards? Will bringing up this issue now cause Kelly to lose the contract? Can she afford to take any risks for her small company?

Step 2: Check facts. Discovery continues with detailed considerations of facts.

Fact: Kelly owns a small firm that is working on a project to update the wetland inventory for the state of Oregon.

Fact: Kelly believes that wetlands exist that are not documented in the inventory. However, the established protocol constrains how Kelly is allowed to add them. This job is financially important for her small business, and Kelly does not want to let it go, as she sees it as a gateway for future project work with the state government.

Fact: The state of Oregon has commissioned Kelly's environmental firm, and she already has begun work.

Fact: A strict protocol exists for establishing wetlands; wetlands can be removed from the database; wetlands can only be added if they already have been mapped.

Fact: Kelly has evidence of additional wetlands, but project resources and protocol prohibit systematic identification.

Fact: Even if the wetlands are verified, there is no established way of mapping them for the project.

Step 3: Identify relevant factors. This step moves into deliberations, as the analysis of situations begins in depth. Do the tasks for the project conflict with the resources and goals for the wetland inventory? The base dataset is a map of probable wetlands (Scranton 2004) that may be limited by the methods used to identify wetlands. The protocol is quite narrow in requiring that added areas already be included in National Wetland Inventory mapping. A conflict between the goals of this data and the probable wetland mapping data can exist. How much additional area is Kelly potentially considering? If the areas are small fringe areas within the accuracy thresholds of existing wetland maps and other data sources for the project, then adding them may be less crucial. However, changing technologies and more accurate data may be grounds for revisiting the protocol. Who can offer guidance on considering revisions to the protocol? Perhaps a revision can be tested while Kelly works on the project.

Step 4: Develop list of options. After deliberating contextual issues, this step creates possible actions.

Option 1: Withdraw from the project.

Option 2: Contact involved agencies.

Option 3: Request a modification to the existing protocol and support to complete the accurate and systematic addition of wetland areas identified during field visits.

Step 5: Test options. This step continues to explore actions by following the systematic use of tests.

Option 1: Withdraw from the project. Harm test: Loss of the contract could threaten the firm's financial solvency and reduce its chances for future state work. Publicity test: Withdrawal could raise touchy questions about the protocol and resulting wetland inventory. Defensibility test: Kelly can face financial difficulties and may lose contacts with the state agency. Reversibility test: Kelly might face challenges legitimating the wetlands inventory if she was working in state government. Colleague test: Colleagues would likely challenge the wisdom of turning down work and losing opportunities to influence future data collection. Professional test: There are numerous potential conflicts with the GISCI's Rules of Conduct. Organization test: Kelly has to keep the business interests of her firm in mind and balance those interests with professional and environmental concerns.

Option 2: Contact involved agencies. Harm test: Agency staff might be thankful for a suggestion, but worried about impacts on the completion time of this project. Publicity test: Delays in completion could have negative consequences for the public. Defensibility test: The additional time in contacting the agencies may add complexities to the project. Reversibility test: If Kelly worked for the state, there may be other channels and possibilities to bring this to the attention of contractors. Colleague test: Colleagues might be thankful in the long run, but worried in the short run about potentially unfunded tasks coming to the project. Professional test: There are numerous potential conflicts with the GISCI's Rules of Conduct. Organization test: Kelly has to keep the business interests of her firm in mind and balance them with professional and environmental concerns.

Option 3: Request a modification to the existing protocol and support to complete the accurate and systematic addition of wetland areas identified during field visits. Harm test: The proposal might be rejected because of the clear scope of work agreed to in the contract, leading to the possible withdrawal of the contract. However, Kelly's expertise could lead to an invitation to future work and revision of the protocol. Publicity test: Finishing the project on time and starting a new project to revise the protocol may be acceptable to the public. Defensibility test: The current project would be completed. Reversibility test: State staff may have some knowledge of the problem and welcome a proposal to revise the protocol. Colleague test: Other GIS professionals may also want a change to the protocol. Professional test: There are no potential conflicts with the GISCI's Rules of Conduct. Kelly has completed the work professionally and has suggested changes. Organization test: The potential for additional project work would be very helpful financially and not undermine Kelly's sense of environmental responsibility.

Step 6: Select choice based on steps 1–5. After the tests, the analysis leads to the selection of Option 3: Request a modification to the existing protocol and support to complete the accurate and systematic addition of wetland areas identified during field visits.

Step 7: Review steps 1–6. It is possible that discovery and deliberation lead to more than one option left in the running. Reviewing the steps can help fine-tune the analysis and develop a presentation of issues that is sound. Variations of the options can be improvements and may replace the option originally chosen in step 6. An approach may also be specified for pursuing an option. For example, first informally contact staff at the state before sending a formal note about issues and ideas about changes to the protocol.

Case Study: Release of Data

This is another case involving discovery and deliberation of issues connected to mapping. The Federal Highway Administration (FWHA) of the U.S. Department of Transportation maintains a national inventory of over 600,000 bridges. States are responsible for conducting periodic inspections of bridges and for reporting their condition to the FHWA. In response to requests from "nongovernmental sources," the FHWA will disclose records from the bridge inventory, but not the locations of individual bridges (which are recorded as latitude and longitude coordinates). Following the collapse of a bridge that caused several fatalities and dozens of injuries, the consultant working with these data for a FHWA project receives a telephone inquiry from a reporter who wishes to map structurally deficient bridges in his state.

Analysis Following the Seven-Step Approach

Step 1: State problem. Stating the problem starts the process of discovery. FHWA data can be provided without locational information about individual bridges. A journalist is requesting the data with locational information.

Step 2: Check facts. Considering facts is part of deliberation.

Fact: The FHWA prohibits the release of locational data from the bridge inventory.

Fact: A bridge in the area collapsed, causing deaths and injuries.

Fact: The consultant has access to the data to complete the FHWA project. No other use is permitted.

Fact: The reporter wishes to make a map of structurally deficient bridges.

Step 3: Identify relevant factors. This step moves into deliberations, as the analysis of situations begins in depth. The key factor appears to be that the consultant has access to the bridge inventory data to complete a project alone. That is, the data cannot be released, based on the terms of the project agreement.

Step 4: Develop list of options. In this step, deliberation moves to develop a list of possible actions.

Option 1: Provide the data to the reporter.

Option 2: Provide a map of bridges with structural problems to the reporter.

Option 3: Retain the data and refer the reporter to the FHWA.

Step 5: Test options. After deliberating contextual issues, this step defines possible actions.

Option 1: Provide the data to the reporter. Harm test: Leads to FHWA canceling the consultant's contract and possible legal action. Publicity test: Public would only get a map. How would the map show the data? Defensibility test: Not defensible under the terms of the contract. Reversibility test: Not good if the consultant is responsible for any of the bridges. Colleague test: Colleagues may support principle but question benefits. Professional test: May lead to conflicts with professional organization. Organization test: This choice would not receive organizational support.

Option 2: Provide a map of bridges with structural problems to the reporter. Harm test: Possible aggravation of the FHWA, even if permitted (which needs to be verified). Publicity test: Public may disregard the map. Defensibility test: Possibly acceptable, needs to be verified. Reversibility test: Any concerns may be eliminated, if FHWA supports the publication of the map. Colleague test: Colleagues may support approach, yet question the effort. Professional test: Involves blurring boundaries between profession and newspaper. Organization test: This choice may receive organizational support, if contract allows for publication of results.

Option 3: Retain the data and refer the reporter to the FHWA. Harm test: Assures that FHWA contract and relationship to consultant are protected. Publicity test: Not possible. Defensibility test: Completely defensible under the terms of the contract. Reversibility test: Reflects value of legal contracts. Colleague test: Colleagues may support principle but question lack of support for this issue. Professional test: May lead to conflicts with professional organization. Organization test: This choice would certainly receive organizational support.

Step 6: Select choice based on steps 1–5. After the tests, the analysis leads to the selection of Option 3: Retain the data and refer the reporter to the FHWA.

Step 7: Review steps 1–6. The review revisits discovery and deliberation activities. The consultant should review the contract and contact the FHWA to determine if the provision of a map would be permitted. Also, the reporter should be referred to FHWA to clarify use of the map and possible acquisition of the data in the future.

REFLECTIONS AND OUTLOOK

These examples are only a starting place for considering the ethical dimensions of professional work. Although it is impossible to anticipate all of the possible situations in which such issues may arise, there are ways we can be better prepared for the times when ethical challenges occur. As Israel and Hay (2009, 167) have pointed out, the greatest barriers to ethical conduct stem from the facts that most of us do not have (1) the philosophical training to help negotiate difficult ethical issues; (2) the ability to recognize ethical challenges as they first appear (rather than after they have escalated into difficult problems); (3) the time to thoroughly consider all issues and options before making decisions; and (4) the talent to anticipate problems before they arise.

However, many organizations and geography programs are trying to remedy this situation, at least in part, by addressing ethics in undergraduate and graduate programs. In the world of GIScience, for instance, ethical training materials have been developed through the gisprofessionalethics.org (2011) project, and ethics are also addressed in the certification procedure of the GIS Certification Institute (2011). Geographers in other fields may find it useful to augment their training with online or traditional courses focusing on applied and professional ethics. Beyond training, discussions of ethics can be an important element of any work environment. So, for example, as new projects are considered, time is set aside to consider explicitly the possible ethical dimensions of the work being planned.

But such reflection is just as important at the individual level—reflecting on ethical issues can be a helpful way to review activities and chart directions for further professional growth and improvement. The seven-step approach presented in this chapter can help bring the ethical dimensions of our work to the fore and open up possibilities for discussions about individual goals and work-related objectives and priorities. Applied ethics can frequently help build a better basis for dealing with the changes we encounter in our careers and work lives. Or, as Jacques Cousteau might observe, such reflection can make us better drivers and help us keep our work and careers on track.

ACKNOWLEDGMENTS

I would like to acknowledge the many colleagues and students who have taken part and commented on various uses of the seven-step approach in workshops and classes. Most of all, I want to recognize the contributions of David DiBiase and Dawn Wright: I had the great pleasure of working with them on the Professional GIS Ethics model curriculum project, supported by the National Science Foundation (Award: 0734824). David DiBiase and colleagues wrote the full version of the wetlands group mapping case. I also would like to acknowledge the two anonymous reviewers of an earlier version of this manuscript and the editors for their helpful comments.

REFERENCES

APA (American Planning Association). 2009. AICP code of ethics and professional conduct. http://www.planning.org/ethics/ethicscode.htm (accessed 24 June 2011).

AAG (Association of American Geographers). 2009. Statement on professional ethics. http://www.aag.org/cs/about_aag/governance/statement_of_professional_ethics (accessed 24 June 2011)

Appiah, K. 2008. *Experiments in ethics*. Cambridge, MA: Harvard University Press.

Davis, M. 1999. *Ethics and the university*. New York, NY: Routledge.

DiBiase, D., M. DeMers, A. Johnson, K. Kemp, A. Taylor Luck, B. Plewe, and E. Wentz, eds. 2006. *Geographic information science and technology body of knowledge*. Washington, DC: Association for American Geographers and University Consortium for GIScience.

DiBiase, D, C. Goranson, F. Harvey, and D. Wright. 2009. The GIS professional ethics project: Practical ethics education for GIS pros. Proceedings of the 24th International Cartography Conference. Santiago, Chile, 15–21 November.

GIS Certification Institute. 2011. Code of ethics. http://www.gisci.org/code_of_ethics.aspx (accessed 24 June 2011).

gisprofessionalethics.org. 2011. Ethics education for geospatial professionals. https://www.e-education.psu.edu/research/projects/gisethicsproducts (accessed 24 June 2011).

Israel, M., and I. Hay. 2009. *Private people, secret places: Ethical research in practice*. In *Aspiring Academics*, eds. M. Solem, K. Foote and J. Monk, 167–178. Upper Saddle River, NJ: Prentice-Hall.

Keefer, M., and K. D. Ashley. 2001. Case-based approaches to professional ethics: A systematic comparison of students' and ethicists' moral reasoning. *Journal of Moral Education* 30 (4): 377–398.

Quinn, M. J. 2006. On teaching ethics inside a computer science department. *Science and Engineering Ethics* 12 (2): 335–343.

Quinn, M. J. 2011. *Ethics for the information age*, 4th ed. Boston, MA: Addison-Wesley.

Scranton, R. 2004. The application of geographic information systems for delineation and classification of tidal wetlands for resource management of Oregon's coastal watersheds. Master's thesis, Marine Resources Management Program, Oregon State University, Corvallis.

Walker, M. U. 1998. Moral understandings. A feminist study in ethics. New York, NY: Routledge.

Correspondence: Department of Geography, University of Minnesota, Minneapolis, MN 55455, e-mail: fharvey@umn.edu.

16

Creating the Life You Want: Lifelong Professional Development for Geographers

Pauline E. Kneale and Larch Maxey

For many students and early-career professionals, thinking about lifelong professional development can seem "too much too soon." Why should one care so soon about lifelong professional development? Skip forward a year or five, however, and you may find yourself wanting to move from, say, the business or government sector to work with a nongovernmental organization (NGO). Suddenly, the value of professional development becomes clear; it can mean the difference between being stuck in an okay job, and creating the life you want.

Geographers have a wide range of knowledge, skills, experience, and interests that place us in an excellent position to create fulfilling lives and careers. Professional development can help us to succeed in the areas that appeal to us and that build on our existing strengths, talents, and interests in professional and personal activities. Consider the following example of how regularly taking time to reflect on your current situation could help you manage your life and career:

> I take a day each summer to get out last year's diary, work plans, and feedback from the annual review by my supervisor. I realize what did and didn't happen, spotting where serendipitous changes have occurred. Some years run quite well to plan, others are more random. I aim to work out when next year's conferences, workshops, and courses will be happening, and to decide which ones to focus on. It's a good time to think about what will just have to be ignored and what will have to be foregrounded. The main question for me is, "What do I need to learn now to keep ahead of [fill in the blank]?" It's harder to think about the next five-year plan, but it's fun and a good way to catch up. It's an away day with yourself. (Pauline)

Professional development can also build one's self-confidence and enthusiasm. As Serge Dedina (profiled with his wife Emily Young in Chapter 9) said to us, "I obtained a grant from a foundation that allowed me to travel with my family, but also finish a book and travel to marine-protected areas in Australia and New Zealand. I came back super excited about my work. That was the best professional development that I could think of."

WHAT IS LIFELONG PROFESSIONAL DEVELOPMENT?

Lifelong professional development (hereafter, professional development) is the creative and innovative building of knowledge, skills, and attitudes throughout a person's life. It means being flexible and responsive to changes as you progress from early career to mid- and late career. Whereas the first chapter in this book by Greiner and Wikle focuses on planning for your initial move into the workforce, we discuss professional development as a process that encompasses the subsequent steps of your career path. As several chapters in this book make clear, personal and professional lives overlap and inform each other continuously. Professional development is a perfect illustration of this.

In this chapter we also focus on organizations and other resources that can support the development of your career, encouraging you to build on these links with all aspects of your life. Explore your opportunities for inclusive, holistic approaches, remembering that different cultures use different nomenclatures: lifelong learning, lifewide learning, permanent education, continuing professional development, continuing professional education, and professional development planning. All of these terms encompass similar ideas.

Embracing professional development offers personal and societal benefits. These benefits may include a greater sense of empowerment, self-confidence, self-respect, and control over your personal situation. You may enhance your ability to create a better work-life balance, more fulfilling relationships, and financial security. Societal benefits include better relationships, greater equality and social mobility, and a more effective, dynamic economy. These aims and benefits have come to be recognized in a number of policy documents by a range of international organizations, including the United Nations (UNESCO 1996), the European Commission (2000, 2002), the World Bank (2003), and the Organization for Economic Cooperation and Development (2004, 2007).

THINK HOLISTICALLY

Professional development is a concept that is being embraced in a number of fields, such as achievement motivation (Weiner 1985; McClelland 1987; Johnson 1990; Elliot and Thrash 2001) and positive psychology (Snyder and Lopez 2002). The learning may be formal (provided by educational institutions and professional organizations, often leading to qualifications), non-formal (typically on-the-job learning that may or may not lead to formalized certificates), and informal (this includes unplanned, often unintentional opportunities that are part of everyday life). The chance nature of professional development embraced within this holistic approach reflects the "happenstance" element Moser and Donelson refer to in Chapter 13 on careers in consulting. Remember that professional development need not be complex or onerous; it can be as simple as keeping up to date with current practice in your field. The focus is on shaping your own planning to fit your current needs and thoughts for the future.

In many businesses, human motivation is a basis for establishing performance-related pay, incentives, bonuses, and promotion schemes. What is important for you to discover is what motivates you. Are you, for example, encouraged or influenced more by satisfaction from voluntary work, the environment and friendliness of your neighborhood, activities with family and friends, your working location, or your current level of job satisfaction? Understanding your own motivation is helpful in focusing on choices. These priorities will change; Judith, for example, loved teaching but noted, "After 18 years teaching high school geography I regrouped, looked at my skills and my love for working for our community. . . . I jumped at the chance to work with the town council, and after two years and courses on law, finance, and people, I am now the Town Clerk, and I am really enjoying it."

Professional development may be workplace-based, employer-led, or employee-led. It may include voluntary and informal learning away from the professional environment. In some organizations, professional development includes formal activities, workshops, and courses that ensure employees are fully aware of particular aspects of their job and are kept abreast of the latest developments. Employee benefits may extend to support for a graduate degree program. As one American professional commented, "I was really enjoying my job, but thinking I could be so much more useful working in Mexico. I started Spanish classes with the main aim of moving south in four to five years, and seven years later it has really paid off. I love it here" (Anonymous).

A plan doesn't always work out, but there are other benefits, as another geographer, Sam, realized: "We decided in our early 20s that we wanted to live in Italy. We both did language classes and researched job options and got really close to moving a couple of times. . . . In the end for family reasons we stayed here, but we have great holidays in Italy and understand opera. The language we learned has been fantastic for enjoying visits more."

If your employer does not offer formal professional development opportunities, it is vital to be proactive and create your own program. In the United States, responsibility for development is increasingly an individual one, whereas in the United Kingdom and Australia, for example, many employers recognize the value of developing their staff and have policies to encourage and support employees, including opportunities for funding or allowing time off. Generally NGOs and smaller businesses are less likely to fund or support development activities, so the trend toward greater NGO employment, particularly of geography graduates, puts the onus on you to be proactive.

ORGANIZATIONS AND OPPORTUNITIES

Informal learning may come through a range of activities, including volunteering in your neighborhood or wider community, or participating in a political organization. It may be paid or unpaid, and it may involve part-time activities or blocks of time nested within employment commitments. Essentially this form of education is characterized by being flexible, embraced by individuals to suit their needs, and likely provided in diverse formats and a range of time scales. Informal learning activities can support your professional development in a variety of ways, providing you with new experiences, contacts, ideas, knowledge, and training. By volunteering with her local youth group, for example, Mandy reports that she gained soft skills such as "confidence and competence in working with younger people." She also gained resources, including "a first aid certificate and glowing references."

For Michael, the choice was to see connections in two areas of his life:

> I did geography at a university which I love. I could have had a music career but decided I wanted to do both geography and music. I have always played the cello in orchestras and practice and play with small groups too. That's really good fun. . . . The skills you get from managing and helping to run voluntary groups like the youth orchestra and concerts help you at work, and work skills help you outside. Having had some experience in an architect's office I decided that's where I'd want to work. On average I do three courses each year, and that's about what the office expects people to do. Some courses are just keeping up-to-date, others are really exciting. . . . What I like is having the balance between different parts of my life.

Some jobs are framed through professional organizations that support the development of their members (Table 16.1). Their structures vary considerably, but all have established criteria for joining initially and in many cases for maintaining membership. This commitment to ongoing personal and professional development maintains the high standards of a profession. Accreditation by professional organizations can demonstrate your competence to people seeking your services. It may be essential where professional indemnity insurance, for example, is dependent on appropriate, recognized professional qualifications.

Exploring the role and activities of relevant professional organizations is a helpful part of researching potential career routes. Organizations seek to maintain high standards of performance for their members. This often involves training programs, workshops, and conferences. Some offer professional insurance and set ethical standards, whereas others are active in

TABLE 16.1 A selection of professional bodies providing accreditation, ethical advice, and training (all websites last accessed 8 August 2011)

United States	
American Planning Association—APA and American Institute of Certified Planners—AICB www.planning.org	Provides leadership nationwide in the certification of professional planners, ethics, professional development, planning education, and the standards of planning practice.
American Society for Photogrammetry and Remote Sensing—ASPRS www.asprs.org/About-Us/What-is-ASPRS.html	Offers professional certification, training, and networking for photogrammetry, remote sensing, geographic information systems (GIS), and supporting technologies
GIS Certification Institute—GISCI www.gisci.org/index.aspx	Provides a comprehensive certification program for geographic information systems (GIS) professionals.
Professional Management Institute—PMI www.pmi.org	Serves practitioners and organizations with standards that describe good practice, globally recognized credentials to certify project management expertise, and resources for professional development, networking, and community.
The Association for GIS Professionals—URISA www.urisa.org	Provides education, certification, and networking for GIS professionals. "A multidisciplinary association where professionals from all parts of the spatial data community come together to share concerns and ideas."
Canada	
Canadian Council of Land Surveyors—CCLS www.ccls-ccag.ca	A national association representing cadastral, geodetic, hydrographic, and photogrammetric surveying, and land information management.
Canadian Institute of Planners—CIP www.cip-icu.ca/web/la/en/default.asp	Offers networking and support and officially recognizes university planning programs. Has reciprocal arrangements with APA, Royal Town Planning Institute, and PIA, and contributed to the formation of the Global Planners Network in 2006. www.globalplannersnetwork.org

Europe

European Water Association—EWA www.ewaonline.de/portale/ewa/ewa.nsf/home?readform	Represents water professionals from 24 organizations in 24 countries. Covers all aspects of the water sector.
European Network of Environmental Professionals—ENEP www.efaep.org	A communication network between environmental professionals in Europe. Provides a knowledge base and information interchange between environmental professionals, public institutions, and private organizations.

Australia

Australasian Housing Institute—AHI www.housinginstitute.org/	Offers awards, development of practice standards/accreditation, a code of practice, professional development events, publications, networks, branches, qualifications, and influences on social housing policy.
Surveying and Spatial Sciences—SSSI Institute www.sssi.org.au	Combines land surveying, engineering, mining surveying, cartography, hydrography, remote sensing, and spatial information science.
UWA Institute of Agriculture—UWA IoA www.ioa.uwa.edu.au/	Coordinates teaching, graduate, and postgraduate training, research, and agribusiness activities with integrated activities related to agriculture, land and water management, rural economy, policy and development, food, and health.

South Africa

AHI—Affordable Housing Institute http://affordablehousinginstitute.org	South African based, global housing finance body. Helps innovators build healthy housing systems worldwide, with an emphasis on the Global South.
South African Planning Institute—SAPI www.sapi.org.za	Provides a forum for all people in planning to debate critical issues affecting planning and development.

UK

Association of Sustainability Practitioners—ASP www.asp-online.org	Offers two levels of membership and CPD for all those in the field of sustainability.
Chartered Institute of Environmental Health—CHEI www.cieh.org/	Professional, awarding, and campaigning body for environmental and public health and safety.
Chartered Institution of Water and Environmental Management—CIWEM www.ciwem.org/	Independent, chartered professional body with an integrated approach to environmental, social, and cultural issues. "Working for the public benefit for a clean, green and sustainable world." UK based with an international membership and remit, offering CPD, training, and lobbying in the broad environmental arena.
Institute of Ecology and Environmental Management—IEE www.ieem.net	CPD and accreditation for members. Advances the science and practice of ecology and environmental management in the UK and internationally.
Royal Institute of Chartered Surveyors—RICS www.rics.org	International organization in the UK, Americas, Australia, New Zealand, and Asia. Offers 40+ courses for those working in land, property, and the built environment, including Assessment of Professional Competency (APC).

campaigning and in political activities in relation to their members' interests, as, for example, the Chartered Institute of Environmental Health (2011) in the United Kingdom. Some, such as the European Network of Environmental Professionals, have adopted a broader international focus (2011). Such bodies allow individual practitioners to work with experts and further develop their own networks, which can be particularly important for individuals running their own companies or working in relatively small organizations.

In the United Kingdom, the Chartered Institution of Water and Environmental Management (CIWEM) advertises itself as "the only independent, chartered professional body and registered charity with an integrated approach to environmental, social and cultural issues" (CIWEM 2011). It focuses on advancing the science and practice of water and environmental management to "sustain the excellence of the people who develop and protect our environment now and for future generations" (CIWEM 2011). As an international organization, it supports the development of individuals through training and professional development. It also works with NGOs, governments, and a range of organizations to shape policy developments at European and United Nations levels.

Also based in the United Kingdom is the Chartered Institute of Housing (CIH 2011), which works in twenty countries drawing members from the academic sector as well as local governments and housing associations. It runs training programs for its staff and for the ongoing development of its members, and it sponsors events to train volunteers, tenants, and those engaged in supporting community housing activities. Working with CIH might be an interesting voluntary activity for an engaged, socially active geographer interested in housing matters in his or her local region.

ADVANCED DEGREES

Other educational opportunities should also be considered. The master's degree is becoming an essential credential for advancement in many careers and parts of the world, including the European Union. In the United States in particular, many people schedule master's and doctoral education into their plans for advancement, as well as a means of transitioning from position to position. The question of when to return to university for further education is difficult to answer because it varies from job to job and from career to career. Discussion with colleagues and peers can often be helpful to developing plans. Also important is the wide range of certificate programs (at both the undergraduate and graduate levels) that are springing up in geography and other fields, including GIScience, environmental management, and other professions. These offer many of the advantages of graduate degrees, but can often be scheduled more easily around full-time schedules.

The rapid emergence of high-quality distance education and e-learning programs, from both nonprofit and for-profit institutions of continuing and higher education, needs to be factored into professional development. Such programs offer greater flexibility for working professionals than traditional classroom instruction and also serve niche audiences at great distances. Geography has been a leader in this area, particularly in GIScience, with certificate and master's programs offered nationally and internationally through programs such as the Pennsylvania State University's World Campus (2011) and UNIGIS International (2011).

Sometimes professional development will involve only a class or two, say in nonprofit accounting regulations, environmental law, or information technology (IT) system administration. Esri, the maker of ArcMap software, now offers a wide array of learning opportunities from podcasts and videos lasting a few minutes to online short courses that can be completed in a few days or a few weeks. All of these trends point in the direction of more opportunities, greater flexibility, and increased specialization of offerings for professional development in the future.

LIFE EVOLVES

In some careers, for example, in finance, teaching, and planning, continuous professional updating and training is required. If we are employed in our own businesses or in small organizations with limited training budgets, or if corporate professional development support is not available, managing our own professional development becomes extremely important. As Hannah's story indicates, your career might not always be entirely self-determined, and redundancy can be an opportunity:

> Hannah was laid off three times in her first five years of employment. "Not my fault. The companies I joined were reorganizing and there was some downsizing. I had two years on a graduate scheme with a water company. That finished with one day's notice. Then two years in human resources and was doing some weekend and evening courses too, all going well until we were taken over and the whole site closed down in a week. . . . I decided press officer work would suit me well. This meant some more courses. . . . I have had this job for six years and three promotions. The layoffs were tough but not my fault. I got experience in diverse businesses that was really helpful later. No regrets. What you don't realize while at a university is that you go on learning at work."

Professional development is too valuable to ignore or leave only to an employer, no matter how enlightened or benevolent. It represents an opportunity to think through *your* priorities, goals, and dreams, to review and change these goals as they evolve throughout your life, and to set about the practical task of achieving them, starting today. Of course, you should be realistic. As one employer says, "in my nonprofit I have had two employees return to grad school expecting me to accommodate them [fund and provide time off]—but due to the economic constraints we couldn't. One finished a night and weekend MBA in nonprofit management—using our nonprofit to fuel his research and work projects. The other quit because we wouldn't accommodate him."

PLANNING IN A CHANGING WORLD

You are already involved in formal, nonformal, and informal professional development, whether or not you have consciously acknowledged it. Excellent learning can come from informal activities that contribute by stealth to your knowledge and skills. Professional development can help you recognize, identify, and value the learning, skills, and attributes acquired from activities in different parts of your life and enable you to be more precise and comprehensive in completing job applications and annual review documents. It's a process that helps you to recognize your existing strengths and qualities and to articulate them clearly. The internship you did last year, the student group you helped to lead, your interest in trail hiking—all of these and many other experiences provide rich informal professional development.

Actively engaging with your own planning allows you to identify gaps or areas you'd like to develop further and to establish a clear plan for achieving your targets. It enables you to make the most of all your experiences, skills, and knowledge, wherever and whenever they occur. Furthermore, a focus on professional development can help to secure progression within an organization and assist in the search for new employment. In the accompanying profile, Nancy Davis Lewis shares some important perspectives on these ideas and the value of professional development throughout her career.

PROFILE 16.1

Nancy Davis Lewis, Ph.D., Research Program Director, The East-West Center (Honolulu, Hawaii)

As a Ph.D. candidate at UC-Berkeley, Nancy Davis Lewis built a small boat and received a Fulbright grant to embark on a two-year sailing journey across the Pacific with her husband and six-year-old son to complete her dissertation research. This irrepressible sense of adventure and **willingness to embrace risks** in the pursuit of new professional horizons has played a major role in shaping her career trajectory.

Before joining the East-West Center in 2002, Nancy had been considering a competing offer of a deanship at a California university. At the time, she was Associate Dean of the College of Social Sciences at the University of Hawaii. She initially ventured into administration in order to learn more about higher education, develop new skills, and help ensure active faculty involvement in campus leadership. While she didn't have any formal training to prepare for this role, Nancy engaged in a lot of self-study and reflection on educational administration and discovered that she had a natural knack for the work. The dean's position would have been a logical next step on the career path she was forging. However, Nancy is a firm believer in the value of lifelong professional development. "In today's market, job seekers need a toolkit of options and I wanted a bigger toolbox," she explains, adding that she sought **greater flexibility** in her personal and professional options for the second half of her career. In addition, becoming a dean would have required her to give up her research in Asia and the Pacific, while the position at the East-West Center has expanded her abilities to explore her scholarly interests: "The position has actually empowered me to add questions [to the Center's research agenda] that are intellectually important to me."

The East-West Center is an independent, public, nonprofit organization whose mission is "to promote better relations and understanding among the people and nations of the United States, Asia, and the Pacific through cooperative study, research, and dialogue" (http://www.eastwestcenter.org). Nancy observes that there are inherent similarities between the work she's doing today and some of the responsibilities she had in previous positions. For example, she still has one foot in the door in academia, serving as an adjunct faculty member at the University of Hawaii, guest lecturing, and advising graduate students, in addition to continuing to explore her long-term interests in issues related to the human dimensions of global change, medical geography, interdisciplinary collaboration, international human rights, and women in the sciences. One major difference between working in a large state bureaucracy and working for a nonprofit organization is that Nancy feels she is able to effect change more readily, enjoying more flexibility and creativity in her role as Research Director and encountering less red tape. In June 2011, Nancy was elected president of the regional, interdisciplinary Pacific Science Association (www.pacificscience.org).

Nancy advises young geographers to "**fill your toolbox** with as many tools and experiences as possible." She is living proof that professional growth and development can arise from formal and informal experiences as well as from academic and nonacademic ones. **Taking advantage of these opportunities** is a personal responsibility; she adds: "You don't teach leadership, but you give people opportunities to lead."

—JOY ADAMS

Ongoing professional development and aspirations need to fit within your broader career management strategy. A willingness to review and evaluate your progress is crucial; every six months is ideal. Life is increasingly competitive. Short-term contracts are more common. Your own circumstances, interests, networks, opportunities, personal pressures, and commitments will change and combine with wider economic uncertainties and fluctuations. As a geographer, you can expect to see the impacts of climate change and natural disasters well into the future. Professional development can help you view these challenges as opportunities to apply your geography abilities and to work toward solutions and change.

In career planning, the establishment of personal aims for the next two weeks, six months, five years, and ten years is emphasized. It is a positive strategy designed to help you develop a personal understanding of your work location preferences, salary aspirations, and needs for a work-life balance. The processes by which you develop your career and professional intelligence, seeking out what is new and exploring training and enhancement opportunities, will not only develop your own career, but potentially allow you to move to a new career. Your career plans, however, may be affected by changing technologies, economics, or political situations that can be hard to predict. Thus the ability to recognize one's skills and attitudes and their relevance in entirely different businesses can be absolutely vital for your continued employment when circumstances evolve rapidly. Your skills may be highly specific to your current job, but in a changing employment climate it is your transferable or soft skills that may make you most employable. Consider the following anecdote:

> Jim reports graduating in 1997: "three years training with a large international company . . . gave me business insights and a few contacts. . . . I was training the under seven's soccer team and wanted to work more locally. So I moved to a much smaller business and really enjoyed the home and community. After about five years, seeing the scope of what was possible, I took some night school programs on leadership, and that made me rethink where I wanted to be. I have had five jobs in fifteen plus years, all good experiences."

It is likely that you will adopt a range of different attitudes to your career (Monk and Jocoy 2009). Managing one's career will always involve a mixture of serendipity, "things that just happen," and planning. Planning and reflection can help you make the most of this combination. When you are "in the right place at the right time," spot a job advertisement, meet someone, or find your personal and family circumstances changing, your professional approach to your development can help you to make the most of it. For example, Jan, a planner in Ohio, kept up her interest in environmental issues throughout her first five years of work in local government. This interest, her involvement with a local campaign group, and a course on using the media allowed her to become the policy spokesperson for a major NGO.

Employability has been understood differently around the world (Rooney et al. 2006). At its simplest level it may be considered as "the capacity to secure and keep employment'" (European Commission 2002, 5). It is now a core goal of many governments and of formal education courses. It is a significant reason to commit to your own professional development as this can significantly enhance your own employability now and throughout your lifetime. If you care for a relative, for example, or need to move closer to your partner, this will shape your career options (Jan Monk discusses these and related issues in Chapter 14.) Life changes and evolves. Being active in your own professional development helps you maximize new situations and see opportunities.

TOOLS FOR PROFESSIONAL DEVELOPMENT

A range of available models can help you to frame your priorities and consider what you value in your life and career. They are particularly useful for helping you to look at the big picture, but they can also help you to see particular tasks and choices more clearly and from different perspectives. There are many resources out there, so don't feel restricted to the following options.

- Double-Loop Learning: This strategy, elaborated by Argyris and Schön (1996), can help you get the most from your professional development by becoming a "reflexive learner"—one who questions the assumptions and ideas on which to base priorities and goals, even as one goes about setting and amending them.
- Balanced View: This program offers an approach to leadership, personal and group organizing, and management that you can apply to all aspects of your professional development and personal life. This simple and practical training uses your own direct experience to develop insight as well as mental and emotional stability in diverse situations. The program is supported by hundreds of video, audio, and book downloads that can be accessed at any time and at no cost from www.balancedview.org (last accessed 15 August 2011).
- Systems Thinking: This process encourages a holistic approach to seeing things in the big picture, emphasizing links and thus helping to ensure that significant parts of your existing experiences, skills, and knowledge, such as informal learning activities, do not get overlooked. It is particularly well suited to geography graduates because this type of thinking is central in understanding many contemporary geographical issues such as sustainability. Try Goal Setting with Systems Thinking (van Zyl, 2010) first.
- Role-Play: Use role playing to put yourself in other positions and to think through opportunities from different perspectives. How would you view your next move if you were twenty years older, if you had a terminal illness, or if you had just won the lottery? Role-play the "job I want." This resource can give you some valuable insights.
- Career Anchors: Edgar Schein identified eight themes or "anchors" around which we typically build our career. Understanding career anchors can help shape your thinking. For a practical introduction, see Schein (2011).
- Goal Setting: This kind of tool can help you broaden and develop the goals that you set for yourself. A useful example is SMART/SMARTER analysis designed to ensure that your goals are Specific, Measurable, Attainable, Relevant, and Time-bound and that you then go on to Evaluate and Re-evaluate them. For a useful introduction to this method, see www.oma.ku.edu/soar/smartgoals.pdf (last accessed 15 August, 2011).
- SWOT Analysis: SWOT (Strengths, Weaknesses, Opportunities, and Threats) analysis can be a very useful precursor to goal setting by helping you identify strengths, weaknesses, opportunities, and threats. Although SWOT began as a tool to guide companies in their strategic direction it is also adaptable for individual professional development and planning. See Chapter 10 for a detailed example of SWOT analysis in a small-business context.
- Timelines, Planners, and Calendars: You can use these resources to look back on how far you have come as well as plan where you would like to get to and when you will take certain actions: whether applying for courses or jobs, taking vacations, and so forth. Plotting vacations and events on an annual calendar gives a shape to your year. Gaps and pressure points show up.
- Backcasting: This planning tool takes the opposite approach to forecasting. With backcasting, you begin in the future—you set objectives and goals for where you wish to be at particular points (five years, ten years) and then you work your way backwards identifying the various steps you need to take to achieve your ambition. Backcasting has been used within strategic

planning for organizations, but it can also be useful for individuals. It is often included as part of a Systems Thinking approach (see, for example, Holmberg and Robèrt 2000).

- Brainstorming: Ideally, this is done in groups for the purpose of generating as many ideas as possible. Do not censor yourself or try to organize or evaluate your ideas while brainstorming. Stay focused on the topic. Having created a list, you now need to organize and develop your ideas. Use friends, family, and colleagues so that you can benefit from the experience, ideas, and perspectives of others. For an introduction to brainstorming and many other free professional development tools, see www.mindtools.com (last accessed 15 August 2011).
- Mentor, Counselor, Coach: These relationships are typically based on face-to-face or mixed/blended sessions that occur over an extended period, from several weeks to several years. The relationships vary in their emphases, but usually focus on helping people identify goals, ideas, and plans. Mentoring is usually free, but you generally have to pay for counseling or coaching. Wilson and Gislason (2010) is one starting point to learn more.
- Supervisor: It may be possible for you to gain your supervisor's direct support for your professional development. The supervisor is a resource and opportunity that many people overlook. Talk to your supervisor about your aspirations.
- Career Advisers: These unsung heroes are found at universities and local governments offering free help, advice, and guidance. They often can help you identify more tailored support that will match your needs.

SUMMARY

Professional development is a valuable process that creates time for you to look at what is truly important to you in life, to pursue interests that you want to develop, and to access the tools and resources to enable you to achieve your goals. Ultimately, thinking, planning, and reflecting can help ensure greater life satisfaction and fulfillment.

Personal measurements of success can include financial rewards, appreciation from others, happiness, and contentment. We suggested earlier that reflection should be on a monthly, half yearly, and yearly basis. Some people enjoy reflecting regularly on their activities through diaries, blogs, wiki sites, or journals. Whatever your choice, keeping track of successes and achievements, and noticing when life swerves you into alternative routes, will be helpful in evaluating current and potential activities. Probably the most valuable skill is being open to new opportunities and trying new approaches.

Planning, developing, and evaluating your ideas and revisiting them regularly should be fun. Use your professional colleagues and others to broaden your insights. Make and create time for yourself to reflect on your deeper values and aspirations. Aim to enjoy the process of learning and reflection, rather than getting overly goal-oriented. Planning can help you remain present and fully aware of all that you have to offer and all that is going on in your life, rather than getting drawn into living in the future only, resenting the current challenges you face. The process can help you develop and clarify your goals, dreams, and plans, but ground them firmly in the present; this is after all the only moment you can truly influence.

Actively reviewing your approach to work and life in general is about taking personal control. It should help you develop self-confidence while improving your coping and self-sufficiency skills. Upward mobility or advancement at work may be an additional benefit. Taking active control of your development should provide positive benefits for coping in all situations, an increased sense of self-respect, and personal strength.

It is your life; let lifelong professional development opportunities help you to enrich it.

REFERENCES

Argyris, C., and D. Schön. 1996. *Organizational learning II: Theory, method and practice*. Reading, MA: Addison-Wesley.

Chartered Institute of Environmental Health. 2011. Home page. www.cieh.org (last accessed 22 August 2011).

Chartered Institute of Housing (CIH). 2011. The CIH home page. www.cih.org (last accessed 22 August 2011).

Chartered Institute of Water Engineers and Managers (CIWEM). 2011. www.ciwem.org.uk/about.aspx (last accessed 22 August 2011).

Elliot, A. J., and T. M. Thrash. 2001. Achievement goals and the hierarchical model of achievement motivation. *Educational Psychology Review 13* (2): 139–156. www.springerlink.com/index/K148523HR271H426.pdf (last accessed February 2011).

European Commission. 2000. *A memorandum on lifelong Learning*. Commission Staff Working Paper Brussels, 30.10.2000 SEC(2000) 1832. www.bologna-berlin2003.de/pdf/MemorandumEng.pdf (last accessed 27 April 2011).

European Commission. 2002. *European report on quality indicators of lifelong learning: Fifteen quality indicators*. www.bologna-berlin2003.de/pdf/Report.pdf (last accessed 27 April 2011).

European Network of Environmental Professionals. 2011. http://www.efaep.org (last accessed 22 August 2011).

Holmberg, J., and K-H. Robèrt. 2000. Backcasting from non-overlapping sustainability principles—A framework for strategic planning. *International Journal of Sustainable Development and World Ecology* 7: 291–308.

Johnson, B. 1990. Toward a multidimensional model of entrepreneurship: The case of achievement motivation and the entrepreneur. *Entrepreneurship Theory and Practice* 14 (3): 39–54.

McClelland, D. C. 1987. *Human motivation*. Cambridge, UK: Cambridge University Press.

Monk, J., and C. Jocoy. 2009. Career planning: Personal goals and professional contexts. In *Aspiring academics: A resource book for graduate students and early career faculty, ed*. Association of American Geographers, 16–31. New York, NY: Prentice Hall.

OECD. 2004. *Lifelong learning*. Policy brief, February. Paris: OECD.

OECD. 2007. *Qualifications and lifelong learning*. Policy Brief, April. Paris: OECD.

Pennsylvania State University's World Campus. 2011. Penn State Online. http://www.worldcampus.psu.edu/ (last accessed 29 September 2011)

Rooney, P., P. Kneale, B. Gambini, A. Keiffer, B. Van Drasek, and S. Gedye. 2006. Variations in international understandings of employability for geography. *Journal of Geography in Higher Education* 30 (1): 133–144.

Schein, E. 2011. Career Anchors online. www.careeranchorsonline.com/SCA/ESabout.do?open=es (last accessed 22 August 2011).

Snyder, C., and J. Lopez. 2002. *Handbook of positive psychology*. Oxford, UK: Oxford University Press.

UNESCO, 1996. *Report to UNESCO of the International Commission on Education for the Twenty-first Century*. Paris, UNESCO. www.unesco.org/education/pdf/15_62.pdf (last accessed 12 August 2011).

UNIGIS International. 2011. Distance Learning GIS Programmes. http://www.unigis.org/ (last accessed 29 September 2011).

Van Zyl, L. 2010. Goal setting with systems thinking. www.buzzle.com/articles/goal-setting-with-systems-thinking.html (last accessed 15 August 2011).

Weiner, B. 1985. An attributional theory of achievement motivation and emotion. *Psychological Review* 92 (4): 548–573.

Wilson, J., and M. Gislason, 2010. *Coaching skills for nonprofit managers and eaders*. San Francisco, CA: Jossey-Bass.

World Bank. 2003. *Lifelong learning in the global knowledge economy: Challenges for developing countries*. Washington, DC: World Bank.

INDEX

H

I

Q

R

S

PHOTO CREDITS

p. 4: Photo courtesy of Valerie Moye; p. 7: Photo courtesy of Carmen Tedesco; p. 18: Photo courtesy of Steve Fearn; p. 20: Photo courtesy of Jason Welborn; p. 29: Photo courtesy of Anne Miglarese; p. 33: Photo courtesy of Sonia Arbona; p. 44: Photo courtesy of Willie Shubert; p. 56: Photo courtesy of Mark Barnes; p. 63: Photo courtesy of Ken Turnbull; p. 68: Photo courtesy of Glenn Letham; p. 77: Photo courtesy of Dan Haug; p. 81: Photo coutresy of Dmitry Messen; p. 88: Photo courtesy of Dave Selkowitz; p. 90: Photo courtesy of Michael Ratcliffe; p. 94: Wikipedia Commons; p. 100: Wikipedia Commons; p. 101: Photo courtesy of Andrew Telfer; p. 103: Photo courtesy of Esther Ofori; p. 105: Wikipedia Commons; p. 113: Photo courtesy of Jane Daniels; p. 116: Photo courtesy of Kate Pearson; p. 120: Photo courtesy of Serge Dedina, Ph.D.; p. 124: Photo courtesy of Kristen Carney; p. 124: Photo courtesy of Anthony Morales; p. 128: Photo courtesy of Ashok Wadwani; p. 137: Photo courtesy of Michelle Kooy; p. 140: Photo courtesy of Reena Patel; p. 149: Photo courtesy of Paula Ann Trevino; p. 156: Photo courtesy of Tara Gettig; p. 178: Photo courtesy of the EPA; p. 181: Photo courtesy of Jeff Young; p. 189: Photo courtesy of Tom Dwyer; p. 205: Photo courtesy of Nancy Davis Lewis, Ph.D.

PHOTO CREDITS